Philipp Zumstein

Extremal Colorings and Extremal Satisfiability

Philipp Zumstein

Extremal Colorings and Extremal Satisfiability

An Interplay between Combinatorics and Complexity Theory

Südwestdeutscher Verlag für Hochschulschriften

Impressum/Imprint (nur für Deutschland/ only for Germany)
Bibliografische Information der Deutschen Nationalbibliothek: Die Deutsche Nationalbibliothek verzeichnet diese Publikation in der Deutschen Nationalbibliografie; detaillierte bibliografische Daten sind im Internet über http://dnb.d-nb.de abrufbar.

Alle in diesem Buch genannten Marken und Produktnamen unterliegen warenzeichen-, marken- oder patentrechtlichem Schutz bzw. sind Warenzeichen oder eingetragene Warenzeichen der jeweiligen Inhaber. Die Wiedergabe von Marken, Produktnamen, Gebrauchsnamen, Handelsnamen, Warenbezeichnungen u.s.w. in diesem Werk berechtigt auch ohne besondere Kennzeichnung nicht zu der Annahme, dass solche Namen im Sinne der Warenzeichen- und Markenschutzgesetzgebung als frei zu betrachten wären und daher von jedermann benutzt werden dürften.

Verlag: Südwestdeutscher Verlag für Hochschulschriften Aktiengesellschaft & Co. KG
Dudweiler Landstr. 99, 66123 Saarbrücken, Deutschland
Telefon +49 681 37 20 271-1, Telefax +49 681 37 20 271-0
Email: info@svh-verlag.de
Zugl.: Zürich, ETH, Diss., 2009

Herstellung in Deutschland:
Schaltungsdienst Lange o.H.G., Berlin
Books on Demand GmbH, Norderstedt
Reha GmbH, Saarbrücken
Amazon Distribution GmbH, Leipzig
ISBN: 978-3-8381-1411-8

Imprint (only for USA, GB)
Bibliographic information published by the Deutsche Nationalbibliothek: The Deutsche Nationalbibliothek lists this publication in the Deutsche Nationalbibliografie; detailed bibliographic data are available in the Internet at http://dnb.d-nb.de.

Any brand names and product names mentioned in this book are subject to trademark, brand or patent protection and are trademarks or registered trademarks of their respective holders. The use of brand names, product names, common names, trade names, product descriptions etc. even without a particular marking in this works is in no way to be construed to mean that such names may be regarded as unrestricted in respect of trademark and brand protection legislation and could thus be used by anyone.

Publisher: Südwestdeutscher Verlag für Hochschulschriften Aktiengesellschaft & Co. KG
Dudweiler Landstr. 99, 66123 Saarbrücken, Germany
Phone +49 681 37 20 271-1, Fax +49 681 37 20 271-0
Email: info@svh-verlag.de

Printed in the U.S.A.
Printed in the U.K. by (see last page)
ISBN: 978-3-8381-1411-8

Copyright © 2010 by the author and Südwestdeutscher Verlag für Hochschulschriften Aktiengesellschaft & Co. KG and licensors
All rights reserved. Saarbrücken 2010

Preface

This book is a postprint of my dissertion submitted to the Swiss federal institute of technology in Zurich (Diss. ETH No. 18603). I would like to thank my supervisors Prof. Dr. Emo Welzl and Prof. Dr. Tibor Szabó for all their advices and encouragements, my colleagues for joint work, all persons who read parts of this thesis and gave me useful comments.

Lungern, March 2010 $\hspace{5cm}$ Philipp Zumstein

Abstract

Combinatorial problems are often easy to state and hard to solve. A whole bunch of graph coloring problems falls into this class as well as the satisfiability problem. The classical coloring problems consider colorings of objects such that two objects which are in a relation receive different colors, e.g., proper vertex-colorings, proper edge-colorings, or proper face-colorings of plane graphs.

A generalization is to color the objects such that some predefined patterns are not monochromatic. Ramsey theory deals with questions under what conditions such colorings can occur. A more restrictive version of colorings forces some substructures to be polychromatic, i.e., to receive all colors used in the coloring at least once. Also a true-false-assignment to the boolean variables of a formula can be seen as a 2-coloring of the literals where there are restrictions that complementary literals receive different colors.

Mostly, the hardness of such problems is been made explicit by proving that they are NP-hard. This indicates that there might be no simple characterization of all solvable instances. Extremal questions then become quite handy, because they do not aim at a complete characteriziation, but rather focus on one parameter and ask for its minimum or maximum value.

The goal of this thesis is to demonstrate this general way on different problems in the area of graph colorings and satisfiability of boolean formulas.

First, we consider graphs where all edge-2-colorings contain a monochromatic copy of some fixed graph H. Such graphs are called H-Ramsey graphs and we concentrate on their minimum degree. Its minimization is the question we are going to answer for H being a biregular bipartite graph, a forest, or a bipartite graph where the size of both partite sets are equal.

Second, vertex-colorings of plane multigraphs are studied such that each face is polychromatic. A natural parameter to upper bound the number of colors which can be used in such a coloring is the size g of the smallest face. We show that every graph can be polychromatically colored with $\lfloor \frac{3g-5}{4} \rfloor$ colors and there are examples for which this bound is almost tight.

Third, we consider a variant of the satisfiability problem where only some (not necessarily all) assignments are allowed. A natural way to choose such a set of allowed assignments is to use a context-free language. If in addition the number of all allowed assignments of length n is lower bounded by $\Omega(\alpha^n)$ for some $\alpha > 1$, then this restricted satisfiability problem will be shown to be NP-hard. Otherwise, there are only polynomially many allowed assignments and the restricted satisfiability problem is proven to be polynomially solvable.

Contents

Preface		i
Abstract		iii
Contents		v
1	**Introduction**	**1**
	1.1 Extremal Ramsey Theory	2
	1.2 Polychromatic Colorings	6
	1.3 Extremal Satisfiability	9
	1.4 Notation	12
2	**Extremal Ramsey Theory**	**17**
	2.1 Bipartite Graphs	20
	2.2 Disjoint Union of Graphs	27
	2.2.1 Cliques	29
	2.2.2 General Graphs	33
	2.3 No upper bound for s in terms of δ	37
	2.4 Cliques—More Colors	41
	2.4.1 Upper Bound	42
	2.4.2 Lower Bounds	51
3	**Polychromatic Colorings**	**59**
	3.1 Polychromatic Edge-Colorings	60

	3.2	Colorings of Plane Multigraphs	67
		3.2.1 The Lower Bound	71
		3.2.2 The Upper Bound	73
	3.3	Special Cases of Plane Graphs	75
		3.3.1 Triangulations	75
		3.3.2 Graphs with Only Even Faces	75
		3.3.3 Outerplanar Graphs	76
	3.4	Connection to Guarding Problems	78
	3.5	Complexity Results for Plane Graphs	79
4	**Extremal Satisfiability**		**91**
	4.1	Problem Description	93
	4.2	Some Observations	96
	4.3	S-SAT and the VC-dimension	98
	4.4	S-SAT and Polynomial Circuits	102
	4.5	S-SAT for Context-Free Languages S	104
	4.6	VC-Dimension of Regular Languages	108
	4.7	Some S-SAT which is not NP-hard	111

Bibliography   **117**

> [Combinatorial mathematics] rendered many services to both pure and applied mathematics. Then along came the prince of computer science with its many mathematical problems and needs—and it was combinatorics that best fitted the glass slipper held out.
>
> S. Jukna

Chapter 1

Introduction

Extremal graph theory and extremal set theory as well as extremal combinatorics in general are beautiful areas of mathematics with connections to fields like probability theory, (linear) algebra, topology, geometry, theoretical computer science. The books [55, 11] give a good overview of these topics. In this thesis we focus on extremal graph colorings and extremal satisfiability.

The term "extremal" means that some *parameter* is maximized or minimized under certain *restrictions*. For example, the classical question in Ramsey theory concerns the minimum number n of vertices needed such that every graph on n vertices either contains a clique of size k or an independent set of size k. Moreover, there are fundamental extremal questions related to complexity, for instance, what is the smallest k such that the k-coloring problem is NP-hard?

This thesis contains three rather independent parts which can be read separately although some connections do exist. Notation common in all chapters is introduced in Section 1.4. Chapter 2 deals with extremal Ramsey theory, which is the study of special colorings of graphs. Chapter 3 is devoted to polychromatic edge-colorings and polychromatic colorings of plane graphs. Chapter 4 contains variants of the satisfiability problem and investigations about their complexity.

Graph coloring is an active and rich field, where the books by Jensen, Toft [54] and by Soifer [84] give an extended overview. In addition, [44] is the standard book in Ramsey theory written by Graham, Rothschild, and Spencer. The philosophy of Ramsey theory is that some structure can always be found within a huge collection of objects. The simplest statement is the pigeonhole principle, a concrete example of which is the fact that in a class with at least 27 students there are always two students whose last name starts with the same letter.

The satisfiability problem is prominent in various areas including logic, artificial intelligence, combinatorial optimization, program and system verification. Satisfiability plays a major role in complexity theory because it was used countless times to deduce NP-hardness of natural problems. Recently, a handbook about satisfiability containing more than 900 pages was published [10].

We proceed by giving a short introduction for each part of this thesis and outlining what will be revealed in the forthcoming chapters.

1.1 Extremal Ramsey Theory

A graph G is called H-*Ramsey*, denoted by $G \to H$, if in every edge-coloring of G with colors red and blue there is a monochromatic copy of H. Furthermore, if every proper subgraph of an H-Ramsey graph G is *not* H-Ramsey, then we say that G is H-*minimal*. We denote the family of all H-Ramsey graphs by $\mathcal{R}(H)$ and the family of H-minimal

1.1. Extremal Ramsey Theory

graphs by $\mathcal{M}(H)$.

The classical theorem of Ramsey implies that for all graphs H the family $\mathcal{R}(H)$ is nonempty and so is $\mathcal{M}(H)$. This was first discovered by Ramsey and published posthumously in [75]. Erdős and Szekeres [34] proved the theorem independently and applied it to a problem in discrete geometry. A significant portion of Ramsey theory is concerned with finding the extremal value of various graph parameters over the family $\mathcal{R}(H)$ or $\mathcal{M}(H)$. The most widely investigated among these questions is the minimization of $n(G)$, the number of vertices, over all graphs $G \in \mathcal{R}(H)$, which gives rise to the classical *Ramsey number* $r(H)$. We use $r(k) = r(K_k)$, where K_k is the k-clique.

The result $r(3) = 6$ is folklore and $r(4) = 18$ is proven in [45]. The exact value of $r(5)$ is already unknown and the best bounds nowadays are $43 \leq r(5) \leq 49$, which are proven in [35, 65]. The growth of the Ramsey number is exponential but the lower and upper bound are still far apart. Erdős proved in [32] that there exists a constant c such that $r(k) \geq ck2^{k/2}$, see also Spencer [85] who improved on the constant c. The currently best known upper bound is proven by Conlon [25] and states that there exists a constant C such that $r(k) \leq k^{-C \frac{\log k}{\log \log k}} \binom{2k}{k}$. To limit the scope we do not mention other results about Ramsey numbers.

Determining $r(H)$ is equivalent to calculating the minimum n such that $K_n \in \mathcal{R}(H)$. Noncomplete Ramsey theory studies graphs other than cliques in $\mathcal{R}(H)$ or $\mathcal{M}(H)$ and their graph parameters. Our main interest in Chapter 2 is the quantity

$$s(H) := \min_{G \in \mathcal{M}(H)} \delta(G),$$

where $\delta(G)$ is the *minimum degree* of the graph G. The parameter $s(H)$ captures the minimum influence of a vertex needed to be important in a H-Ramsey graph. Clearly, $s(H) \geq \delta(H)$. Burr, Erdős, and Lovász [19] introduced $s(H)$ and studied it for H being a clique. The exact value is known for cliques and complete bipartite graphs [19, 21, 40], and a simple lower bound $s(H) \geq 2\delta(H) - 1$ (see Proposition 2.2) is proven by Fox and Lin in [40]. We will extend these results in various ways.

Example. Let $H = C_4$ be the 4-cycle. The simple lower bound yields $s(C_4) \geq 3$ while it is known that K_6 and $K_{5,5}$ are C_4-Ramsey graphs [42, 8]. Any subgraph of these two graphs has minimum degree at most 5 and an H-Ramsey graph always has an H-minimal subgraph. Therefore, $s(C_4) \leq 5$, but this is not optimal. We claim that $G = K_{3,9} \in \mathcal{R}(C_4)$. Let A, B be the partite sets with $|A| = 3, |B| = 9$ and color the edges of G with red and blue. Every vertex $b \in V(B)$ has three incident edges and there are $2^3 = 8$ possible color patterns of these edges. Since $|B| = 9$, the pigeonhole principle implies that there are two vertices $b_1, b_2 \in V(B)$ with the same color pattern. The color blue or red appears at least twice in this color pattern, which yields a monochromatic C_4 containing the vertices b_1, b_2. This shows that $K_{3,9} \in \mathcal{R}(C_4)$. Every subgraph of $K_{3,9}$, in particular any C_4-minimal subgraph, has minimum degree at most 3 which proves that $s(C_4) \leq 3$. Actually, it is also possible to prove that already $K_{3,7}$ is C_4-Ramsey, even C_4-minimal. A generalization of this argument for complete bipartite graphs can be found in [40].

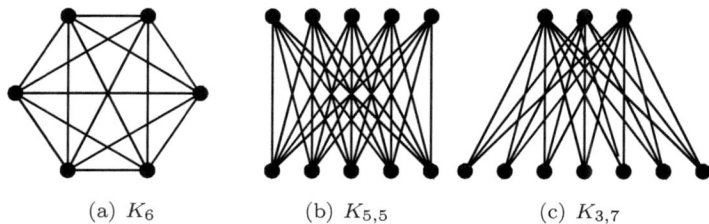

(a) K_6 (b) $K_{5,5}$ (c) $K_{3,7}$

Figure 1.1: C_4-Ramsey graphs

In Section 2.1 we will determine the s-value for a large class of bipartite graphs. This class includes all even cycles, forests, bi-regular graphs, and all connected bipartite graphs with partite sets of the same size. Actually, all these graphs fulfill $s(H) = 2\delta(H) - 1$, i.e., the simple lower bound is tight for them. We will continue in Section 2.2 by ex-

1.1. Extremal Ramsey Theory

ploring the behavior of the s-parameter under taking disjoint union of graphs. Especially, we will prove that a large clique completely dictates the s-value in the disjoint union with a small clique, while a small complete bipartite graph determines the s-value in the disjoint union with a larger complete bipartite graph. For complete graphs as well as for all bipartite graphs considered in Section 2.1, the s-value is always upper bounded by a function of the minimum degree. However, this is not true in general, as we will prove in Section 2.3. Finally, in Section 2.4 we will discuss asymmetric cases and generalizations to more than two colors for cliques.

For proving upper bounds on $s(H)$ one has to show that there exists an H-minimal graph with small minimum degree. We usually show this by explicit constructions. In general, it is not easy to explicitly construct an H-Ramsey graph. The minimality makes such constructions even more difficult. Our usual approach is to proceed in two steps. In the first step, we find a graph G that is not H-Ramsey but in all colorings without a monochromatic copy of H, which are also called critical colorings, a special coloring of some subgraph can be forced. In the second step, we extend G by adding a new vertex (or maybe more than one) for some t-subsets of the vertices $V(G)$ and connect it to all members of this t-set. If everything fits together nicely, we can force a coloring to appear in every critical coloring of G such that no matter how we color the newly introduced edges there is a monochromatic copy of H. This proves that the extended graph is H-Ramsey. Moreover, if we delete all newly introduced vertices, then we would obtain G again, which was by assumption not H-Ramsey. Hence, there exists some graph between G and its extension which is H-minimal. The minimum degree of this graph cannot be larger than t, showing that $s(H) \leq t$.

Chapter 2 is based on the joint work with Tibor Szabó and Stefanie Zürcher [88].

1.2 Polychromatic Colorings

The *art gallery problem* is a famous problem in computer science and it originates from a real-world problem. Imagine an art gallery consisting of one large room which we will think of as a simple polygon P. The task is to place guards at its vertices to make sure they see the whole polygon P. What is the minimum number of guards needed for that? If P is convex then one guard is sufficient, but usually there are some reflex corners and a guard cannot see around such corners. Chvátal [24] proved that there is always a set of at most $n/3$ vertex guards who guard a polygon P with n vertices. Furthermore, this bound is tight. Although Chvátal's paper contains only three pages, Fisk [38] found an even shorter proof of this fact which we will present here: Triangulate

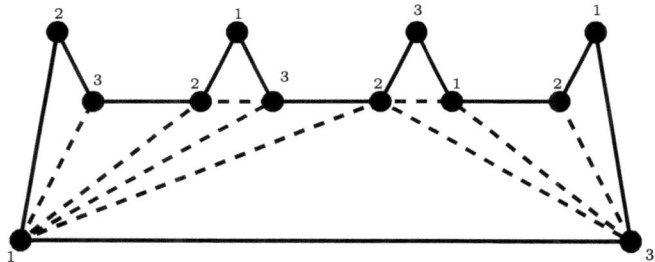

Figure 1.2: A polygon P with a triangulation and a vertex-3-coloring which can be guarded by 4 vertex-guards but not by 3.

P by only adding straight edges inside P. The resulting graph is still outerplanar and therefore can be properly colored with 3 colors. The smallest color class C contains no more than $n/3$ vertices. Every triangle contains a vertex of each color class. We place at each vertex of C a guard and claim that they together guard the whole polygon P. Every point p inside P belongs to some triangle T and one of the vertices of T is in C, say x. Since a triangle is convex the guard at x sees p which proves that the whole polygon P is guarded with at most $n/3$ guards.

The crucial point in the argument above is that every triangle re-

1.2. Polychromatic Colorings

ceives all three colors. A vertex-k-coloring of a plane multigraph G is *polychromatic* if every face receives all colors on its boundary. In contrast to the classical coloring problem in graph theory, it is harder to provide a polychromatic coloring with many colors than with few. Polychromatic colorings correspond to a *combinatorial variant* of the art gallery problem: The input is a plane multigraph G and a vertex guard sees all the faces incident to it. This means especially that we forget about the whole geometry and allow guards also to see around corners if it is still in the same face.

We consider plane *multigraphs* where edges are drawn by any curve connecting the endpoints (see Figure 1.3). This setting is more general and most of our results in Chapter 3 fit into it.

There are several proofs that every plane multigraph without faces of size 1 or 2 can be polychromatically 2-colored (Theorem 3.15). Therefore any plane multigraph on n vertices with no faces of size 1 or 2 can be guarded by $\lfloor \frac{n}{2} \rfloor$ guards. In the combinatorial setting as described above this is tight.

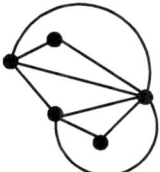 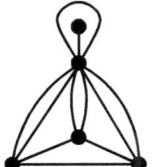 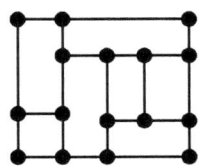

(a) triangulation with multiedges   (b) plane multigraph with 2-faces   (c) non-degenerate rectangular subdivision

Figure 1.3: Plane Multigraphs

From a result of Hoffmann and Kriegel [50] it follows that any plane, bipartite, 2-connected simple graph is polychromatically 3-colorable (see Theorem 3.26). Horev and Krakovski [53] showed that any connected plane graph G without faces of size 1 or 2 and maximum degree at most 3, which is not K_4 or a subdivision of K_4 on 5 vertices, is polychromatically 3-colorable. In [52] it is shown that every bipartite cubic plane graph has a polychromatic 4-coloring (with the possible exception of the outer face).

A subdivision of a rectangle into rectangles is called rectangular subdivision and it is non-degenerate if no four rectangles meet in a point. Dinitz et al. [31] showed that it is possible to color the vertices of any non-degenerate rectangular subdivision S with three colors such that each rectangle in S has at least one vertex of each color. They conjectured that this is also possible with four colors. And indeed, a proof by Guenin [46] of a conjecture by Seymour [82] concerning the edge-coloring of a special class of planar graphs, directly implies such a 4-coloring [30]. Keszegh [58] investigates polychromatic colorings of so-called n-dimensional guillotine-partitions.

For a plane multigraph G, let $g(G)$ be the smallest number of vertices any face has. There cannot be a polychromatic coloring of G with more than $g(G)$ colors. In [53] it was asked if there exists a constant c such that every plane multigraph G has a polychromatic coloring with $g(G) - c$ colors. We will show that this is not true. It is proven in Section 3.2 that for $g \geq 5$ every plane multigraph with $g(G) = g$ there exists a polychromatic coloring with $\lfloor \frac{3g-5}{4} \rfloor$ colors. On the other hand we construct plane simple graphs G with $g(G) = g$ where every polychromatic coloring can use at most $\lfloor \frac{3g+1}{4} \rfloor$ colors. The main steps for proving that a multigraph is polychromatically k-colorable are: (1) we assign to each face almost all its vertices such that a vertex is not assigned more than twice; and (2) we consider special edge-colorings which are called polychromatic edge-colorings. Actually, we will first consider these special edge-colorings in Section 3.1 and after that we show how to apply them to obtain polychromatic colorings of plane multigraphs in Section 3.2.

In Section 3.3 we consider special cases of plane graphs. Triangulations are plane multigraphs such that every face is a 3-cycle (Figure 1.3(a)). By the above discussion we know that every triangulation is polychromatically 2-colorable and sometimes it is also polychromatically 3-colorable. A triangulation is polychromatically 3-colorable if and only if it is properly 3-colorable. For example, a plane embedding of K_4 is *not* polychromatically 3-colorable. We will explore this connection in more details. Furthermore, we study multigraphs with even faces only and outerplanar multigraphs. It will be proven that any outerplanar multigraph G with $g = g(G) \geq 3$ is polychromatically g-colorable.

Section 3.4 explains the connection to guarding problems in more details. Finally, complexity questions are considered in Section 3.5. The decision problem whether a plane multigraph is polychromatically k-colorable is in P for $k = 2$ and it is NP-complete for $k = 3$ or $k = 4$. Moreover, we continue by giving some more restrictive decision problems. For a set L of integers we consider the decision problem whether a plane multigraph with faces of sizes only in L is polychromatically 3-colorable. There is an almost complete characterization shown for which sets L the problem is in P and for which it is NP-complete.

Chapter 3 is based on joint work with Noga Alon, Robert Berke, Kevin Buchin, Maike Buchin, Péter Csorba, Saswata Shannigrahi, and Bettina Speckmann [5].

1.3 Extremal Satisfiability

A *boolean formula* is a well-formed expression containing boolean variables, the logical AND, the logical OR, and the logical negation. A boolean formula f over the variables $V = \{v_1, \ldots, v_n\}$ is *satisfiable* if there is an assignment in $\{\text{true}, \text{false}\}^V$ such that f evaluates to true. Satisfiability is the problem to decide whether a given formula is satisfiable, and it was the first problem which was proven to be NP-complete [26, 63].

There are two main questions which guide us from here: What special restriction on the SAT problem can guarantee that the decision problem is trivial, i.e., the answer is always YES or always NO? What restriction on SAT are possible such that it remains NP-hard?

There are three ways to restrict the SAT problem:

(i) Restrict on special formulas f,
(ii) change the satisfying condition, or
(iii) restrict the solution space $\{\text{true}, \text{false}\}^V$.

The most common restrictions are of the form (i) and we will mention here some results (not including monotone, planar, or linear SAT). It is well-known that every boolean formula has an equivalent boolean formula in conjunctive normal form (CNF). The SAT problem restricted to CNF formulas where all clauses contain k literals, denoted by k-SAT, is NP-hard for $k \geq 3$ and it is polynomial time solvable for $k = 2$.

Next, restrictions on the number of occurrences in a k-CNF formula are discussed. It is shown in [61] that if there exists some unsatisfiable k-CNF formula where every variable occurs only s times, then the restricted satisfiability problem is NP-hard. Define $f(k)$ to be the largest integer s such that all k-CNF formulas with variables not occurring more than s times are satisfiable. This is a very interesting extremal parameter. An application of the Lovász Local Lemma shows that $f(k) \geq \frac{2^k}{ek}$ [61]. The very recent construction in [41] shows that this is tight up to a constant factor, i.e., $f(k) \in \Theta(\frac{2^k}{k})$. Also recently, an algorithmic version of the Lovász Local Lemma has been established which implies that for k-CNF formulas with variables that occur at most $\frac{2^{k-5}}{k}$ times not only the decision problem can be solved but also a satisfying assignment can be found in polynomial time [68].

These results are linked to the dependencies of clauses: Two clauses have a conflict (negative dependency) if there is a variable which occurs in one positively and in the other negatively, and they have a positive dependency if they share a literal. One might expect that an unsatisfiable CNF formula should contain many conflicts, which was the starting point of the investigations in [81], where it is shown that for k large enough $2.69^k \leq c_k \leq 3.55^k$, where c_k denotes the minimum number of conflicts in an unsatisfiable k-CNF formula.

1.3. Extremal Satisfiability

The second approach (ii) was investigated by Schaefer [78]. His setup allows to take formulas f which are conjunctions of logical relations. A logical relation is a subset of all possible assignments and therefore it is a generalization of the disjunction in the conjunctive normal form. He gave a complete characterization of the classes of relations leading to polynomial time algorithms, and the other classes are NP-hard. This dichotomy result is astonishing because one could expect that there would also be intermediate cases, that are neither in P nor NP-hard. It follows from Schaefer's theorem that the variants NAE-SAT or exactly-one SAT are NP-hard. The recent paper [3] considers a refinement with respect to subtler complexity classes.

We investigate the third way (iii) by restricting the search space. Normally, every assignment to the variables is allowed, but we want to forbid some assignments a priori. Given a set $S \subseteq \{0,1\}^*$ of assignments, the S-SAT problem asks whether for a formula F over n variables there is an assignment $S_n := S \cap \{0,1\}^n$ that satisfies F. If so, then F is called S-*satisfiable*. If $|S_n|$ is polynomial in n and S_n can be enumerated in polynomial time then S-SAT is in P. To exclude this case we concentrate on *asymptotically exponential* families, for which there exists some $\alpha > 1$ such that $|S_n| \in \Omega(\alpha^n)$. In fact, we will work with a generalization of asymptotically exponential families in Chapter 4.

The question whether S-SAT is NP-hard for all asymptotically exponential S was first stated by Cooper [27]. We will disprove this conjecture by constructing an exponential S such that S-SAT is not NP-hard, provided P $\neq$ NP (Section 4.7)

The S-SAT problem is still hard under different notions of hardness: We show that if S-SAT is in P for some exponential S, then SAT, and thus every problem in NP, has polynomial circuits (Section 4.4). This would imply that the polynomial hierarchy collapses to its second level [56]. Since this is widely believed to be false, it is a strong indication that S-SAT is a hard problem in general.

A natural way to describe a language S is by a grammar (if there exists one). Therefore, we will go further and concentrate on families S_n given by a regular or context-free grammar. In both cases, the S-SAT problem turns out to be NP-hard for every exponential family S. The main tool to prove NP-hardness of S-SAT is to compute large index sets for which every assignment can be realized by S_n (see Section 4.3). The maximum size of such an index set is the VC-dimension of S_n. It is hard to compute the VC-dimension in general. Moreover the size of S_n is large, and therefore this approach seems not applicable for a polynomial reduction. However, if S is given by a finite deterministic state machine then we can compute the VC-dimension and an index set of this size in linear time (Section 4.6). Even if S is given by a context-free grammar, we can compute large index sets shattered by S_n (not necessarily a maximum one), which will lead to NP-hardness proofs of such S-SAT (see Section 4.5).

Chapter 4 is based on joint work Dominik Scheder [80].

1.4 Notation

We denote by $\mathbb{N}$ the set of *natural numbers* $1, 2, 3, \ldots$, and $\mathbb{N}_0 = \mathbb{N} \cup \{0\}$. We use the notation $[n] := \{1, 2, \ldots, n\}$ and the set of all k-subsets of a set S is denoted by $\binom{S}{k}$. The set of all functions $f : V \to W$ is denoted by W^V.

Asymptotics. Let f, g, h be real positive functions. We write $f \in O(g)$ if there exist n_0, c such that $f(n) \leq cg(n)$ for all $n \geq n_0$. Normally, for exponential functions we neglect polynomial factors and write $f \in O^*(g)$ instead. Moreover, we use the notation $h \in \Omega(g)$ if $g \in O(h)$.

Multigraphs. A *multigraph* G is a pair consisting of a finite set $V(G)$ of *vertices* and a multiset $E(G)$ of *edges* from the set $\binom{V(G)}{2} \cup \binom{V(G)}{1}$. We use $n(G) = |V(G)|$ and $e(G) = |E(G)|$. For an edge $e \in E(G)$, the elements in e are called its *endpoints*. A *loop* is an edge $e \in E(G)$ with

1.4. Notation

only one endpoint. *Multiple edges* are edges with the same endpoints. A *graph* G is a multigraph without loops and without multiple edges, i.e., the edges are a subset of $\binom{V(G)}{2}$. If we want to emphasize that G is a graph rather than a multigraph, then we also say that G is a *simple graph*. We assume in the following that multigraphs have *no loops* if not otherwise stated.

Multigraphs without loops. Two vertices $u, v \in V(G)$ are *adjacent* in the multigraph G if $\{u, v\} \in E(G)$. The *neighborhood* of a vertex v in the multigraph G is denoted by $N_G(v)$ and contains all vertices adjacent to v, the *degree* of v $\deg_G(v)$ equals to the number of edges incident to v. If the multigraph G is clear from the context, we usually just write $N(v)$ and $\deg(v)$. The *minimum degree* of a multigraph G is denoted by $\delta(G)$, and the *maximum degree* by $\Delta(G)$. A vertex $v \in V(G)$ is isolated if $N_G(v) = \emptyset$.

A *vertex-k-coloring* of G is a map $\varphi : V(G) \to \{1, \ldots, k\}$ and an *edge-k-coloring* of G is a map $\varphi : E(G) \to \{1, \ldots, k\}$. A vertex-$k$-coloring φ is *proper* if for every edge $\{u, v\} \in E(G)$, $\varphi(u) \neq \varphi(v)$. An edge-$k$-coloring is *proper* if for every vertex $v \in V(G)$, all edges incident to v have different colors. The *chromatic number* $\chi(G)$ is the smallest k such that there exists a proper vertex-k-coloring of G. A multigraph G is *bipartite* if $\chi(G) \leq 2$.

Plane multigraphs. A *drawing* of a multigraph G is a function defined on $V(G) \cup E(G)$ that assigns to each vertex v a point $f(v)$ in the plane and assigns each edge with endpoints u, v a curve connecting $f(u), f(v)$. A *plane embedding* of G is a drawing of G such that no two curves meet in a point other than a common endpoint. Note that we do not require that the curves are straight line segments. A *plane multigraph* is a multigraph G together with a plane embedding of G. We denote the set of faces of G by $F(G)$. For a plane multigraph G, a *dual graph* G^* is a plane multigraph which has for each face $f \in F(G)$

a vertex $x_f \in V(G^*)$ drawn inside f. An edge $e \in E(G)$ with face a on one side and face b on the other side gives rise to an edge $e^* \in E(G^*)$ connecting x_a and x_b. The dual graph G^* can contain loops and multi-edges also if G itself is a simple graph (Figure 1.4).

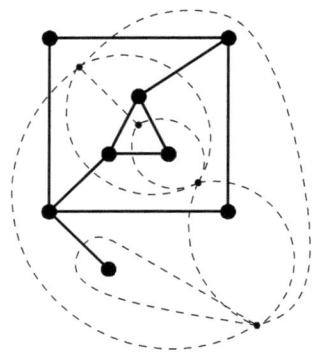

Figure 1.4: A simple plane graph and its dual graph.

Simple graphs. For $A, B \subseteq V(G)$ let $E(A, B)$ denote the set of edges with one endpoint in A and the other one in B. For $A = B$, we abbreviate $E(A) := E(A, A)$. For a graph G and a set $S \subseteq V(G)$ we denote its induced subgraph by $G[S]$, i.e.,

$$G[S] = \left(S, E(G) \cap \binom{S}{2}\right).$$

We write $G - U$ for the graph $G[V(G) \setminus U]$. The *independence number* $\alpha(G)$ of G is the largest size of a set $S \subseteq V(G)$ such that $G[S]$ contains no edges.

A graph is called *2-connected* (*2-edge-connected*) if after the deletion of any vertex (edge) the graph is still connected.

We say that there is a *copy* of H in G if there is an injective map $\varphi : V(H) \to V(G)$ such that if $\{h_1, h_2\} \in E(H)$ then also $\{\varphi(h_1), \varphi(h_2)\} \in E(G)$. An injective map $\varphi : V(H) \to V(G)$ such that $\{h_1, h_2\} \in E(H)$ if and only if $\{\varphi(h_1), \varphi(h_2)\} \in E(G)$ is called an *induced embedding* of

1.4. Notation

H in G. If an induced embedding of H in G is also bijective, then we say that G and H are *isomorphic* and write $G \cong H$. The clique number of G, denoted by $\omega(G)$, is the largest t such that there is a copy of K_t in G.

For two graphs H_1, H_2 let H_1', H_2' be isomorphic copies such that $H_i' \cong H_i$ for $i = 1, 2$ and $V(H_1') \cap V(H_2') = \emptyset$. Then $H_1 + H_2$ denotes the *disjoint sum* of H_1 and H_2, with $V(H_1 + H_2) = V(H_1') \cup V(H_2'), E(H_1 + H_2) = E(H_1') \cup E(H_2')$. Furthermore, tH denotes the disjoint sum $H + H + \ldots + H$ of t isomorphic copies of H. The *join* $H_1 \vee H_2$ is the graph obtained from $H_1 + H_2$ by adding all edges $\{x, y\}$ where $x \in V(H_1')$ and $y \in V(H_2')$.

Directed multigraph. An *edge-orientation* of a multigraph G is a map $\varphi : E(G) \to V(G)$ such that the image of each edge e is one of its endpoints and we say that the edge e points towards the vertex $\varphi(e)$. A *directed multigraph* is a graph with an edge-orientation. For a directed multigraph G and a vertex $v \in V(G)$, the *in-degree* $d_G^-(v)$ is the number of edges pointing towards v and $d_G^+(v) = \deg_G(v) - d_G^-(v)$ is the out-degree of v.

Ramsey theory. The hypergraph Ramsey number $r_k(a_1, \ldots, a_c)$ is the smallest number $n \in \mathbb{N}$ such that for every c-coloring of the k-subsets of $[n]$ there is an $i \in [c]$ and an a_i-subset $A \subseteq [n]$ such that all elements of $\binom{A}{k}$ are colored with the ith color. We write $r_k(a)$ if all a_i are equal to a and for $k = 2$ we omit the index k and just write $r(a_1, \ldots, a_c)$. We talk about a *symmetric* case if all a_i are the same, and otherwise we refer to an *asymmetric* case.

Boolean formulas. A *boolean variable* is a variable with values true and false which we also interpret as integers 1 and 0. The negation of a variable v is denoted by $\bar{v}$ or $\neg v$ and it evaluates to $1 - v$. The logical AND is denoted by $\wedge$ and the logical OR by $\vee$ and their evaluation is as usual. A *boolean formula* F is a well-formed syntactic expression containing boolean variables as well as $\vee, \wedge, \neg$, and parantheses. Denote by $\mathrm{vbl}(F)$ all the variables which occur at least once in F. The *size* of a boolean formula is the length of the expression.

A *literal* is a variable v or its negation $\bar{v}$. A *k-clause* is a disjunction of exactly k literals not containing the same literal twice or a variable and its negation. For example, $v_3 \vee \bar{v_5} \vee v_9$ is a 3-clause. A formula is in *conjunctive normal form* (CNF) if it is a conjunction of clauses, furthermore f is a *k-CNF formula* if it is a formula in conjunctive normal form with k-clauses only.

An *assignment* to the variables $V = \{v_1, \ldots, v_n\}$ is a function in $\{\text{true}, \text{false}\}^V$ or $\{0,1\}^V$ and it evaluates a boolean formula on the variables V in the usual way.

A boolean formula over variables V is satisfiable if there exists an assignment in $\{\text{true}, \text{false}\}^V$ such that f evaluates to true under this assignment.

Languages. Let Σ be a finite *alphabet* (normally $\Sigma = \{0,1\}$). The *empty word* has length 0 and is denoted by ε. For $n \in \mathbb{N}_0$, the set of all words over Σ of length n will be denoted by Σ^n and it consists of all concatenations (sequences) of n elements of Σ. Moreover, the set of all words is $\Sigma^* = \bigcup_{n \in \mathbb{N}_0} \Sigma^n$. A *language* is any subset of Σ^*. For two languages L_1, L_2, we denote by $L_1 L_2$ the language containing all the words w of the form $w = w_1 w_2$ for $w_1 \in L_1, w_2 \in L_2$. For a language L we define inductively $L^0 = \{\varepsilon\}$ and $L^{n+1} = L^n L$ for $n \in \mathbb{N}_0$. Moreover, we define $L^* = \bigcup_{n \in \mathbb{N}_0} L^n$.

> Party mathematics is an important tool in the repertoire of the socially gifted mathematician, and one of the all-time favorite stories tell us that at a party of six people there are at least three people who know each other, or three people who do not know each other. As mathematicians started to get invited to larger parties, they began working on the general case.
>
> M. Schaefer

Chapter 2

Extremal Ramsey Theory

Noncomplete Ramsey theory—the term was introduced by Burr [15]—considers colorings of *noncomplete* graphs. For complete graphs one graph parameter, as for example the number of vertices, determines all other graph parameters. This is not the case for noncomplete graphs where there is a whole bunch of graph parameters which are of independent interest.

A graph G is called H-*Ramsey*, denoted by $G \to H$, if in every edge-coloring of G with colors red and blue there is a monochromatic H. Furthermore, if every proper subgraph G' of an H-Ramsey graph

G is *not* H-Ramsey, then we say that G is H-*minimal*. We denote the family of all H-Ramsey graphs by $\mathcal{R}(H)$ and the family of H-minimal graphs by $\mathcal{M}(H)$.

The classical theorem of Ramsey states that for all graphs H the family $\mathcal{R}(H)$ is nonempty, and therefore also $\mathcal{M}(H)$ is nonempty. The quantity $r(H) = \min_{G \in \mathcal{R}(H)} n(G)$ is the classical *Ramsey number* of H (for a regularly updated survey on Ramsey numbers of all kinds of graphs, see [74]).

One of the first results in the area of noncomplete Ramsey theory states that for every H there exists G with the same clique number $\omega(G) = \omega(H)$ and $G \to H$, see [39, 70].

Instead of minimizing the clique number, we can also ask how small the chromatic number of G can be such that still $G \to H$. Denote by $r_\chi(H)$ the minimum chromatic number of all H-Ramsey graphs. This parameter is characterized for all graphs H in [19], although in most cases its actual value is not known. Some new definitions are needed to understand that result. For a set $\mathcal{G}$ of graphs we denote by $r(\mathcal{G})$ the minimum n such that in every red/blue edge-coloring of K_n there is a monochromatic copy of *some* graph in $\mathcal{G}$. Furthermore, the set of images of all homomorphisms from H is denoted by $\hom(H)$. It is proven that $r_\chi(H)$ is equal to $r(\hom(H))$, see [19]. In particular, it is easy to see that for H bipartite $r_\chi(H) = 2$, i.e., for every bipartite graph H there exists a bipartite graph G such that $G \to H$. Another special case is for $H = K_k$ where it is easy to see that $\hom(K_k) = \{K_k\}$ and therefore $r_\chi(K_k) = r(K_k)$. The greedy coloring of a graph G with maximum degree Δ shows that $\chi(G) \leq \Delta + 1$ and therefore it follows that every K_k-Ramsey graph has a vertex of degree at least $r(K_k) - 1$.

Another natural parameter is the *size Ramsey number* $\hat{r}(H)$, which is the minimum number of the edges over all graphs in $\mathcal{R}(H)$. The size Ramsey number was introduced by Erdős, Faudree, Rousseau, and Schelp in [33] and studied more extensively by many others (see [37] for a recent survey). Clearly, the number of edges in the complete graph with $r(H)$ vertices is an upper bound for the size Ramsey number. It is interesting to note that this bound is tight for $H = K_k$:

Theorem 2.1 ([33]). *For a positive integer m we have*
$$\hat{r}(K_k) = \binom{r(K_k)}{2}.$$

Proof. Let G be a K_k-Ramsey graph and let G' be a χ-critical subgraph of G, i.e., $\chi(G') = \chi(G) \geq r(K_k)$ and for every $x \in V(G') : \chi(G' - x) < \chi(G)$. Then it is easy to see that $\delta(G') \geq \chi(G') - 1$ and $n(G') \geq \chi(G')$. Therefore for the number of edges
$$e(G) \geq e(G') \geq \binom{\chi(G')}{2} \geq \binom{r(K_k)}{2}. \qquad \square$$

There are results about Ramsey-minimal graphs considering whether $\mathcal{M}(H)$ is finite or infinite for a given graph H. The following characterization is known: $\mathcal{M}(H) < \infty$ if and only if H is the disjoint union of a (possibly empty) matching and at most one star with an odd number of edges, see [19, 76, 69, 17, 18, 36].

For the parameters like minimum degree or connectivity one has to restrict to Ramsey-minimal graphs to obtain reasonable questions. Our main interest in this chapter is in the quantity
$$s(H) := \min_{G \in \mathcal{M}(H)} \delta(G),$$

i.e., the minimum of the minimum degrees over all H-minimal graphs. This parameter was introduced and first studied by Burr, Erdős, and Lovász [19]. By the minimality condition one cannot just simply add vertices of small degree to an H-Ramsey graph and thereby get a small upper bound for $s(H)$. On the contrary, each of the vertices, in particular a vertex of minimum degree, has to be *important* to produce a monochromatic copy of H.

Proposition 2.2 ("simple lower bound", [40]). *For all graphs H*
$$s(H) \geq 2\delta(H) - 1.$$

Proof. Assume for contradiction that there exists an H-minimal graph G with $\delta(G) < 2\delta(H) - 1$. Let $v \in V(G)$ be a vertex of minimum degree. By the minimality there exists an edge-coloring c of $G - v$ without monochromatic copy of H. We extend c to an edge-coloring c' of G by coloring at most $\delta(H) - 1$ of the edges incident to v red and the remaining at most $\delta(H) - 1$ edges blue. The degree of v in the red graph is at most $\delta(H) - 1$ implying that v cannot be part of a red copy of H. Similarly, v cannot be part of a blue copy of H. Therefore, there is no monochromatic copy of H in the edge-coloring c' of G, which contradicts to the assumption that G is H-Ramsey. □

A clique of size $r(H)$ is H-Ramsey, but maybe it is not H-minimal. By deleting some edges or vertices of the clique (not everything) the minimum degree cannot increase (this holds for every regular graph). Therefore we have $s(H) \leq r(H) - 1$. The determination of $r(K_k)$ is out of reach currently and is one of the most notorious open problems in combinatorics. In a striking contrast, $s(K_k)$ turned out to be more approachable and was computed exactly for every k by Burr et al. [19, 21]: they obtained that $s(K_k) = (k-1)^2$. An alternative proof was found by Fox and Lin [40]. They showed also that Proposition 2.2 is tight for all complete bipartite graphs $K_{a,b}$, i.e., $s(K_{a,b}) = 2\min\{a,b\} - 1$, and they raised the question whether the simple lower bound (2.2) would be tight for any other graph.

2.1 Bipartite Graphs

We answer the above question affirmatively by providing a large class of bipartite graphs which all attain the simple lower bound. This class contains paths, even cycles, and more generally, all trees and all biregular bipartite graphs.

A *bipartition* (A, B) of a bipartite graph H is a partition of the vertices $V(H) = A \cup B$ such that $E(A, B) = E(H)$. If H is connected,

2.1. Bipartite Graphs

then there is only one bipartition. We define the parameters $a(H)$ and $b(H)$ by

$$a(H) := \min\{|S| : (S, V(H) \setminus S) \text{ is a bipartition}\},$$

and $b(H) := n(H) - a(H)$. For a bipartite graph H with bipartition (A, B) let $\Delta_A(H)$ ($\Delta_B(H)$) be the largest among the degrees of vertices in A (B). A bipartite graph H is called *bi-regular* if there is a bipartition (A, B) such that $\deg(x) = \Delta_A(H)$ for every $x \in A$ and $\deg(x) = \Delta_B(H)$ for every $x \in B$.

For all graphs H with $a(H) = a, b(H) = b$ we obviously have $H \subseteq K_{a,b}$, but $H \not\subseteq K_{a-1,m}$ for all $m \in \mathbb{N}$.

Lemma 2.3. *Let H be a bipartite graph with $a(H) = a$. Then for all positive integers m*

$$K_{2a-2,m} \not\to H.$$

Proof. Let us denote by V the partite set of $K_{2a-2,m}$ having size $2a-2$. We partition V into two sets V_1 and V_2, each of size $a-1$, and we color the edges incident to V_1 red and the edges incident to V_2 blue. This edge-coloring does not contain a monochromatic H, since the red and blue graph are each copies of $K_{a-1,m} \not\supseteq H$. □

The edge-coloring in the above proof has the property that both monochromatic subgraphs are copies of $K_{a-1,m}$. We call such an edge-coloring of $K_{2a-2,m}$ *balanced*. For a fixed d, we will show that if an edge-coloring of $K_{2a-2,m}$ has no monochromatic copy of H then it contains a balanced coloring of $K_{2a-2,d}$, provided m is large enough.

Lemma 2.4. *Let H be a bipartite graph with $a(H) = a$ and $b(H) = b$ and let d be an integer. Then there exists an integer $m = m(a, b, d)$ such that in every red/blue edge-coloring of $K_{2a-2,m}$ there exists*

(i) *a monochromatic copy of H, or*
(ii) *a copy of $K_{2a-2,d}$ with a balanced coloring.*

Proof. Clearly, if the statement is true for some d, then it is true for all $d' \leq d$, because a balanced coloring of $K_{2a-2,d}$ contains a $K_{2a-2,d'}$ with a balanced coloring. Hence, we may assume that $d \geq b$. Moreover, it is enough to prove the lemma for $H = K_{a,b}$. Set $m := (d-1) \cdot 2^{2a-2} + 1$ and denote by V and W the partite sets of $K_{2a-2,m}$ of size $2a-2$ and m, respectively. Consider an arbitrary edge-coloring c of $K_{2a-2,m}$ with colors red and blue. Let $v_1, v_2, \ldots, v_{2a-2}$ be the elements of V in some ordering. Assign for each vertex $w \in W$ a vector $p(w) \in \{\text{red}, \text{blue}\}^{2a-2}$ such that $p(w)_i = c(\{w, v_i\})$ for $i = 1, 2, \ldots, 2a-2$. There are 2^{2a-2} possible p-vectors. By the pigeonhole principle there exist at least d vertices $w_1, \ldots w_d \in W$ with the same p-vector. If the number of red entries in $p(w_1)$ is at least a, then the vertices $w_1, \ldots, w_d$ and a of the vertices of V corresponding to the red entries of $p(w_1)$ form a monochromatic red copy of $K_{a,d} \supseteq K_{a,b}$. The case of at least a blue entries in $p(w_1)$ is analogous. Otherwise, $p(w_1)$ has $a-1$ red and $a-1$ blue entries, meaning that the vertices $w_1, \ldots, w_d$ and V induce a $K_{2a-2,d}$ with a balanced coloring. □

We define the bipartite incidence graph $S(m, k) = (A \cup B, E)$ for integers m, k by

$$A = \{1, 2, \ldots, m\}, \quad B = \binom{A}{k}, \quad E = \{\{a, T\} : T \in B, a \in T\}.$$

Lemma 2.5 (Nešetřil, Rödl [70]; cf. Diestel [29] p.264). *Let H be a bipartite graph.*

(i) *There exist integers m, k such that H can be embedded into $S(m, k)$. In fact, we can choose $k = a(H) + 1$.*

(ii) *For every $k, m \in \mathbb{N}$ there exists an integer m' such that*

$$S(m', 2k-1) \to S(m, k).$$

Corollary 2.6. *For every bipartite graph $H = (A \cup B, E)$ we have*

$$s(H) \leq 2a(H) + 1.$$

2.1. Bipartite Graphs

We will modify the above lemma using a slightly different construction and thereby improve the bound on k. We will then apply this to derive our main theorem.

Definition 2.7. *Let G be a graph, $J \subseteq V(G)$, and $k \geq 1, \ell \geq 1$. Then we define $\mathcal{T}_k^\ell(G; J)$ to be the graph (V', E') with*

$$V' = V(G) \cup \left(\binom{J}{k} \times [\ell] \right),$$

$$E' = E(G) \cup \left\{ \{x, (M, i)\} \,:\, M \in \binom{J}{k}, x \in M, i \in [\ell] \right\}.$$

The graph defined above can be obtained from G by first designating a subset J of the vertices of G and then for each k-tuple M of J adding ℓ new distinct vertices and connecting them to all vertices in M. It is clear, that $|V'| = |V(G)| + \binom{|J|}{k} \cdot \ell$, $|E'| = |E(G)| + \binom{|J|}{k} \cdot \ell \cdot k$. Furthermore, note that unless $|J| < k$, the degree of all new introduced vertices is k. Observe that $\mathcal{T}_k^1(E_n; [n]) = S(n, k)$ for E_n being the empty graph on the vertices $[n] = \{1, 2, \ldots, n\}$.

Lemma 2.8. *Let $H = (A \cup B, E)$ be a bipartite graph.*

(i) *There exist n, k, ℓ such that H can be embedded in $\mathcal{T}_k^\ell(E_n, [n])$. In fact, we can choose $k = \Delta_B(H)$ and map A into $V(E_n)$.*

(ii) *For every n, k, ℓ there exists n', ℓ', with the property that*

$$\mathcal{T}_{2k-1}^{\ell'}(E_{n'}, [n']) \to \mathcal{T}_k^\ell(E_n, [n]),$$

such that the set corresponding to $V(E_n)$ in the monochromatic copy of $\mathcal{T}_k^\ell(E_n, [n])$ is contained in $V(E_{n'})$.

Proof. (i) Set $n = |A| + \Delta_B(H)$, $k = \Delta_B(H)$, $\ell = |B|$. In order to find an embedding $\varphi : H \to \mathcal{T}_k^\ell(E_n, [n])$, first arbitrarily map A onto $[|A|]$ then process the vertices of B in an arbitrary order: For each $w \in B$ it holds that $|N(w)| \leq k$, so we can choose $L = \varphi(N(w)) \cup \{|A| + 1, \ldots, |A| + (k - \deg(w))\} \in \binom{[n]}{k}$ and map w to (L, i) for some unused i (there is at least one unused i by the definition of ℓ).

(ii) Set $\ell' = 2\binom{2k-1}{k}(\ell-1)+1$, $n' = r_k(n,n,2k-1)$ (the k-uniform hypergraph Ramsey number for three colors) and let $K = T^{\ell'}_{2k-1}(E_{n'};[n'])$. Color the edges of K with red and blue. The degree of each vertex $(M,i) \in V(K) \setminus [n']$ is $2k-1$, so there is a color $c_{M,i}$ which appears at least k times among the edges incident to (M,i). Hence we can define a function $\varphi : \binom{[n']}{2k-1} \times [\ell'] \to \{\text{red}, \text{blue}\} \times \binom{[n']}{k}$ such that all edges of K between (M,i) and the second component $(\varphi(M,i))_2$ (which is a k-element subset of M) is colored with the first component $(\varphi(M,i))_1$. For any fixed $M \in \binom{[n']}{2k-1}$, there are $2\binom{2k-1}{k}$ many possible φ-values. Thus, by the definition of ℓ' and the pigeonhole principle, at least ℓ of the vertices from $\{(M,1),(M,2),\ldots,(M,\ell')\}$ have the same φ-value; let us denote this value by φ_M.

We now define an auxiliary coloring of the k-tuples $\binom{[n']}{k}$. For a subset $S \in \binom{[n']}{k}$, if there exists an $M \in \binom{[n']}{2k-1}$ such that $(\varphi_M)_2 = S$ then S receives the color $(\varphi_M)_1$ (if there are more than one such M then we choose one of them arbitrarily). This way we obtain a partial red/blue coloring of $\binom{[n']}{k}$ which we extend by giving each yet uncolored k-tuple the color white. By the choice of n' there is

(a) a set of size n with only red k-tuples or

(b) a set of size n with only blue k-tuples or

(c) a set of size $2k-1$ with only white k-tuples.

Case (c) does not occur because by definition, every $2k-1$ tuple $M \in \binom{[n']}{2k-1}$ does contain a red or a blue k-tuple, namely $(\varphi_M)_2$.

The cases (a) and (b) are symmetric, therefore we can assume that we have a set $A' \subseteq [n']$ of size n containing only red k-tuples of the auxiliary coloring. This means that for each k-tuple $T \subseteq A'$, there is a $(2k-1)$-set $M_T \supseteq T$ such that $(\varphi_{M_T})_2 = T$ and $(\varphi_{M_T})_1 = \text{red}$. Hence there are ℓ vertices of the form (M_T,i) each of which has only red edges towards T. In particular there is a red copy of $T^{\ell}_k(E_n;[n])$ in K. □

By part (i) and (ii) of Lemma 2.8, we have the following corollary.

2.1. Bipartite Graphs

Corollary 2.9. *For every bipartite graph* $H = (A \cup B, E)$

$$s(H) \leq 2\min\{\Delta_A(H), \Delta_B(H)\} - 1\,.$$

The following theorem is our main result in this section. It shows that for a large class of bipartite graphs the simple lower bound is tight.

Theorem 2.10. *Let H be a bipartite graph with $\delta(H) \geq 1$ and assume that there exists a bipartition (A, B) of H such that $|\{v \in B : \deg(v) > \delta(H)\}| \leq a(H) - 1$. Then*

$$s(H) = 2 \cdot \delta(H) - 1\,.$$

Proof. Let (A, B) be a bipartition of H with $|\{v \in B : \deg(v) > \delta(H)\}| \leq a(H) - 1$. Let $S \subseteq \{v \in B : \deg(v) = \delta(H)\}$ be an arbitrary subset such that $|B \setminus S| = a(H) - 1 =: a'$. Clearly $S \neq \emptyset$ because there is no bipartition where one part is smaller than $a(H)$. Let $N(S) \subseteq A$ denote the set of vertices adjacent to at least one vertex in S. The graph $H^* = H[S \cup N(S)]$ has a bipartition, namely $(S, N(S))$, such that $\deg(s) = \delta(H), \forall s \in S$, i.e., $\Delta_S(H^*) = \delta(H)$. According to Lemma 2.8 there exist integers $n = n(H^*)$ and $\ell = \ell(H^*)$ with the property that

$$\mathcal{T}^{\ell}_{2\delta(H)-1}(E_n;[n]) \rightarrow H[S \cup N(S)]\,, \tag{2.1}$$

such that in the monochromatic copy of $H[S \cup N(S)]$ the set $N(S)$ is contained in $V(E_n)$. Without loss of generality we can assume that $n \geq |A|$.

By Lemma 2.4, there is an integer $m = m(a(H), b(H), n)$ such that in every edge-coloring of $G = K_{2a',m}$ there exists a monochromatic H or there is a copy of $K_{2a',n}$ with a balanced coloring. Let L and M be the partite sets of G with size $2a'$ and m, respectively. Now we show that

$$\mathcal{T}^{\ell}_{2\delta(H)-1}(G;M) \rightarrow H\,. \tag{2.2}$$

Let c be an arbitrary red/blue edge-coloring of $\mathcal{T}^{\ell}_{2\delta(H)-1}(G;M)$. The restriction of c to $E(G)$ either contains a monochromatic H and we are done, or otherwise there is a copy K of $K_{2a',n}$ with a balanced coloring. Let L and $M' \subseteq M$, $|M'| = n$, be the partite sets inducing K.

Consider $T^\ell_{2\delta(H)-1}(E_n; M')$ which is a subgraph of $T^\ell_{2\delta(H)-1}(G;M)$. By (2.1) there exists a monochromatic, say blue, copy T of $H[S \cup N(S)]$, such that the image of $N(S)$ is contained in M'. Since $|M'| \geq |A|$ we have space to embed the vertices of $A \setminus N(S)$ in $M' \setminus V(T)$. Hence the union of T and the blue copy of $K_{a',n}$ in K contains a blue copy of H and (2.2) follows.

On the other hand by Lemma 2.3 $G \not\to H$, and hence there is an H-minimal graph G', such that $G \subseteq G' \subseteq T^\ell_{2\delta(H)-1}(G;M)$. The minimum degree of G' is clearly at most $2\delta(H) - 1$ and the theorem follows. □

Corollary 2.11. *Let* $k \geq 2$.

(i) *For all paths* P_k, *we have* $s(P_k) = 1$.
(ii) *For all even cycles* $C_{2k}, k \geq 2$, *we have* $s(C_{2k}) = 3$.
(iii) *For all bi-regular bipartite graphs* H *with* $\delta(H) \geq 1$, *we have* $s(H) = 2\delta(H) - 1$.
(iv) *For all connected bipartite graphs* $H = (A \cup B, E)$ *with* $|A| = |B|$ *we have* $s(H) = 2\delta(H) - 1$.
(v) *For every tree* T, *we have* $s(T) = 1$.

Proof. Parts (i)-(iv) are immediate. For (v), let X and Y be the partite sets of the tree T. We can easily apply Theorem 2.10 unless $|X| \neq |Y|$ and all vertices of minimum degree are contained in the larger of the two partite sets.

Hence assume that $|X| > |Y| = a(T)$ and the set of all vertices of degree 1, denoted by S, is contained in X. To apply Theorem 2.10 it is enough to show that $|X \setminus S| < |Y|$. Fix an arbitrary vertex $r \in Y$ as the root of the tree and define the successor relation according to it. All vertices in $X \setminus S$ have at least one successor in Y and these all have to be different (because there are no cycles). Thus the function $\text{succ}: X \setminus S \to Y$ is injective. Since the root vertex r is not the successor of any vertex, we have

$$|Y| \geq |\operatorname{succ}(X \setminus S)| + 1 \geq |X \setminus S| + 1\,.\qquad\square$$

Define $\mathcal{G}_\delta$ to be the family of bipartite graphs H with $\delta(H) = \delta$ for which there is a bipartition (A, B) such that $|\{v \in B : \deg(v) > \delta(H)\}| \leq a(H) - 1$. Theorem 2.10 states that for each graph in $\mathcal{G}_\delta$ we have $s(H) = 2\delta - 1$.

Observation. If $H_1 \in \mathcal{G}_\delta$ and H_2 bipartite with $\delta(H_2) \geq \delta$ then $H_1 + H_2 \in \mathcal{G}_\delta$.

Let (A_1, B_1) be a good bipartition of H_1 and let (A_2, B_2) be a bipartition of H_2 such that $|B_2| = a(H_2)$. We have $a(H_1 + H_2) = a(H_1) + a(H_2)$ and it is easy to see that $(A_1 \cup A_2, B_1 \cup B_2)$ is a good bipartition of $H_1 + H_2$.

The following corollaries are immediate consequences of the above observation and Corollary 2.11.

Corollary 2.12. *For all forests F without isolated vertices, we have $s(F) = 1$.*

Corollary 2.13. *For all bipartite graphs H with $1 \leq \delta(H) \leq \Delta(H) \leq 2$, we have $s(H) = 2\delta(H) - 1$.*

Indeed, the graphs in Corollary 2.13 are disjoint sums of paths and even cycles.

2.2 Disjoint Union of Graphs

Taking the disjoint union of two graphs is arguably the simplest graph operation. The common parameters behave simple under this operation, e.g., the chromatic number satisfies $\chi(G + H) = \max\{\chi(G), \chi(H)\}$, the independence number has the property $\alpha(G + H) = \alpha(G) + \alpha(H)$, and for the clique number $\omega(G + H) = \max\{\omega(G), \omega(H)\}$ holds. What can be said about the Ramsey extremal properties, like the Ramsey number r, or the parameter s introduced before? We will show that their behavior can be much more complex.

Our main considerations here are for cliques and complete bipartite graphs, and our main focus is how a small graph can affect the behavior in the disjoint union with a large graph. Let us first concentrate on complete bipartite graphs. By the discussion from the previous section we have

$$s(K_{b,b} + K_{c,c}) = 2\min\{b, c\} - 1 \,.$$

Thus the smaller complete bipartite graph determines this s-value completely. If $b < c$ then this value is different from $s(K_{c,c})$. Especially, we see that there are $(K_{b,b} + K_{c,c})$-minimal graphs that are not $K_{c,c}$-minimal, i.e., there exist a graph G that is $K_{c,c}$-Ramsey but not $(K_{b,b} + K_{c,c})$- Ramsey. On the other hand we will show that a small clique does not affect the Ramsey parameters in the union with a large clique in a most general way: the K_t-Ramsey graphs and the $(K_t + K_s)$-Ramsey graphs are the same when $s \leq t - 2$. This behavior leads to the following definitions.

Definition 2.14. *Two graphs H and K are* Ramsey-equivalent *if the set of H-Ramsey graphs and the set of K-Ramsey graphs are the same. Otherwise, H and K are called* Ramsey-separable.

The set of all H-Ramsey graphs is monotone, i.e., if J is H-Ramsey then so is every supergraph of J. The minimal elements with respect to the subgraph relation constitute the family $\mathcal{M}(H)$. Thus, if H and K are Ramsey-equivalent, then $\mathcal{M}(H) = \mathcal{M}(K)$ and consequently $s(H) = s(K), r(H) = r(K), \hat{r}(H) = \hat{r}(K)$.

Clearly, the relation of being Ramsey-equivalent is an equivalence relation and we have in addition that if the graphs A, C are Ramsey-equivalent and $A \subseteq B \subseteq C$, then all three graphs A, B, C are Ramsey-equivalent.

2.2.1 Cliques

The smallest clique is K_1 which contains just one point. Up to now, we have always assumed that we do not have isolated vertices. The reason is that we can handle them separately by the following proposition.

Proposition 2.15. *Let H be a graph without isolated vertices and for some $t \geq 1$ define $H' = H + tK_1$.*

(i) *If $t > r(H) - n(H)$ then H and H' are Ramsey-separable, $r(H') = n(H')$, and $s(H') = 0$*
(ii) *If $t \leq r(H) - n(H)$ then H and H' are Ramsey-equivalent, $r(H') = r(H)$, and $s(H') = s(H)$.*

Proof. (i) There exists a graph K on exactly $r(H)$ vertices that is H-minimal, but every H'-Ramsey graph has to contain at least $n(H') > r(H)$ vertices, which implies that K cannot be a H'-Ramsey graph. The graph $K + (t - r(H) + n(H))K_1$ is clearly H'-minimal and has minimum degree 0 and $n(H')$ many vertices.

(ii) We claim that a graph G is H-Ramsey if and only if it is H'-Ramsey. If G is H-Ramsey then, by the definition of $r(H)$, $n(G) \geq r(H)$. Hence in any edge-coloring of G there are at least $r(H) - n(H)$ vertices besides a monochromatic copy of H to accommodate the t isolated vertices of H'. □

The Ramsey parameters for the disjoint union of some clique with a 2-clique is worked out completely in the following theorem.

Theorem 2.16. *For $t \geq 4$*

$$r(K_2 + K_2) = 5 \quad r(K_3 + K_2) = 7 \quad r(K_t + K_2) = r(K_t)$$
$$s(K_2 + K_2) = 1 \quad s(K_3 + K_2) = 1 \quad s(K_t + K_2) = (t-1)^2.$$

Proof. By Lemma 2.2 we have $s(K_r + K_2) \geq 1$ for all $r \geq 2$. A matching of size three shows that $s(K_2 + K_2) = 1$. It is easy to see that K_4 is not $K_2 + K_2$-Ramsey but K_5 is.

For $r = 3$, we claim that $K_6 + K_2$ is $(K_3 + K_2)$-minimal, and thus $s(K_3 + K_2) = 1$.

In any red/blue edge-coloring of K_6 there is *at least one* monochromatic K_3. To avoid a monochromatic $K_3 + K_2$ the edge-coloring must contain two vertex-disjoint monochromatic copies of K_3: one in blue and one in red. No matter how we color the extra edge, we will get a monochromatic $K_3 + K_2$, i.e., $K_6 + K_2 \to K_3 + K_2$.

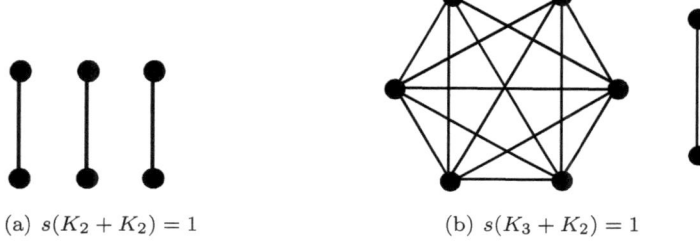

(a) $s(K_2 + K_2) = 1$ (b) $s(K_3 + K_2) = 1$

Figure 2.1: Minimal graphs

The graph K_6 minus one edge has an edge-coloring without a monochromatic K_3, and K_6 is not $(K_3 + K_2)$-Ramsey: The coloring consisting of a red K_4 and all remaining edges blue contains no monochromatic $K_3 + K_2$. Hence $K_6 + K_2$ is $(K_3 + K_2)$-minimal, $s(K_3 + K_2) = \delta(K_6 + K_2) = 1$, and $r(K_3 + K_2) > 6$.

In any red/blue edge-coloring of K_7 there is a monochromatic K_3, say it is red. The edges between the other 4 vertices all have to be colored blue. To avoid a blue $K_3 + K_2$, all the edges between these two parts have to be red, which will yield a red $K_3 + K_2$.

For $t \geq 4$ we apply Theorem 2.17 with $s = 2, a_1 = t$, and $a_2 = 2$, and use that $r(t, t-1) > 2t$ for $t \geq 4$ and the result $s(K_t) = (t-1)^2$. □

Remark. Let $t \geq 4$ and $H = K_t + H_2$. Then H has minimum degree 1 but its s-value grows quadratically in t. This example shows that the trivial lower bound (Lemma 2.2) can be arbitrarily far away from the actual value of $s(H)$.

2.2. Disjoint Union of Graphs

Theorem 2.17. *Let $a_1 \geq a_2 \geq \ldots \geq a_s \geq 1$ and define $H_i := K_{a_1} + \ldots + K_{a_i}$ for $1 \leq i \leq s$. If $r(a_1, a_1 - a_s + 1) > 2(a_1 + \ldots + a_{s-1})$, then H_s and H_{s-1} are Ramsey-equivalent.*

Proof. Since H_{s-1} is a subgraph of H_s, if $G \to H_s$ then also $G \to H_{s-1}$. Thus it suffices to show that $G \to H_{s-1}$ implies $G \to H_s$.

Let G be a graph such that $G \to H_{s-1}$ and suppose for contradiction that $G \not\to H_s$. Let c be a red/blue edge-coloring of G without monochromatic H_s. Without loss of generality, we may assume that there is a blue copy of H_{s-1}, and let S_1 be its vertex set. Since c has no blue H_s, the coloring restricted to $V(G) \setminus S_1$ has no blue K_{a_s}. Define H_0 to be the empty graph. Let i be the largest index such that $V(G) \setminus S_1$ contains a red H_i and let S_2 be its vertex set (it may happen that S_2 is empty). Since c has no red H_s, we have $i < s$. The coloring c restricted to $V(G) \setminus (S_1 \cup S_2)$ contains no red K_{a_1}.

Our goal is now to recolor some of the edges of G such that the resulting coloring c' contains no monochromatic K_{a_1}. We have $|S_1 \cup S_2| = |V(H_{s-1})| + |V(H_i)| \leq 2(a_1 + \ldots + a_{s-1}) < r(a_1, a_1 - a_s + 1)$ so by the definition of the Ramsey number we can recolor the edges inside $S_1 \cup S_2$ such that there is no red K_{a_1} and no blue $K_{a_1 - a_s + 1}$. All edges between $S_1 \cup S_2$ and $V(G) \setminus (S_1 \cup S_2)$ are recolored to blue, while the colors of the other edges do not change. The largest blue clique restricted to $S_1 \cup S_2$ has at most $a_1 - a_s$ vertices, the largest blue clique in $V(G) \setminus (S_1 \cup S_2)$ has at most $a_s - 1$ vertices, which implies that c' contains no blue copy of K_{a_1}. Since there are no red edges between $S_1 \cup S_2$ and $V(G) \setminus (S_1 \cup S_2)$, the largest red clique contains less than a_1 vertices. Therefore there is no monochromatic K_{a_1} in c'. This is a contradiction to $G \to H_{s-1}$ and the proof is complete. □

We can start with $H_1 = K_{a_1}$ and use the above theorem when the condition is satisfied to show that K_{a_1} and $H_2 = K_{a_1} + K_{a_2}$ are Ramsey-equivalent. Then we can continue with H_2 and check the condition again to show that it is Ramsey-equivalent to H_3. Since Ramsey-equivalent is an equivalence relation we can follow that H_1 and H_3 are as well Ramsey-equivalent under these assumptions. By repeating this argument s times we obtained the following corollary.

Corollary 2.18. *Let $a_1 \geq \ldots \geq a_s \geq 1$ be such that*
$$r(a_1, a_1 - a_i + 1) > 2(a_1 + \ldots + a_{i-1}), \quad \forall i = 2, \ldots, s.$$
Then $K_{a_1} + \ldots + K_{a_s}$ is Ramsey-equivalent to K_{a_1}.

Let $s < \frac{r(t, t-k+1) - 2(t-k)}{2k}$ and $k \leq t - 2$. The variable s fulfills $r(t, k+1) > 2(t + (s-1)(t-k))$. According to Corollary 2.18 the graphs $K_t + sK_k$ and K_t are Ramsey-equivalent.

For the two extremes of the spectrum of k, we spell out the concrete bounds by substituting known results for the Ramsey number.

(a) $K_t + sK_{t-2}$ is Ramsey-equivalent to K_t for *some* $s = \Omega\left(\frac{t}{\log t}\right)$,

(b) $K_t + sK_2$ is Ramsey-equivalent to K_t for $t \geq 4$ and *some* $s = \Omega(t2^{t/2})$.

For (a), one uses that $r(t, 3) = \Omega\left(\frac{t^2}{\log t}\right)$ proven by Kim [59], for (b) one can use $r(t, t-1) = \Omega(t2^{t/2})$ proven by Erdős [32]. Naturally, the question arises, whether we can go further, i.e., are the graphs $K_t + K_t$ or $K_t + K_{t-1}$ also Ramsey-equivalent to K_t.

Proposition 2.19. *Let $t \geq 1$.*

(i) *K_t and $K_t + K_t$ are Ramsey-separable.*

(ii) *K_t and K_{t-1} are Ramsey-separable.*

Proof. (i) Let $R = r(K_t, K_t)$ and $G = K_R$. Then $G \to K_t$ but $K_{R-1} \not\to K_t$. Extend an edge-coloring of K_{R-1} without a monochromatic K_t arbitrarily to $K_{R-1} \vee x \cong K_R$. All monochromatic K_t in this extended coloring have to contain the vertex x and therefore we do not find two vertex-disjoint ones. This proves $G \not\to K_t + K_t$.

(ii) Nešetřil and Rödl [70] proved that $\min\{\chi(G) : G \to H\} = \chi(H)$. Thus two graphs with different chromatic number are Ramsey-separable, in particular K_t and K_{t-1} are Ramsey-separable. □

2.2. Disjoint Union of Graphs

It is shown in [20] that $r(tK_3) = 5t, \forall t \geq 2$. This implies that sK_3 and tK_3 are Ramsey-separable for $s \neq t$.

Theorem 2.20 ([20]). *For all $t \in \mathbb{N}$ and graphs G on n vertices*

$$(2n - \alpha(G))t - 1 \leq r(tG) \leq (2n - \alpha(G))t + C(G),$$

where $\alpha(G)$ denotes the maximum size of an independent set in G and $C(G)$ is a constant depending on G. Moreover, there exists $t_0 \in \mathbb{N}$ and $D(G)$ such that for $t \geq t_0$

$$r(tG) = (2n - \alpha(G))t + D(G).$$

Burr [16] worked out the constant $D(G)$ explicitly for complete graphs and cycles: for t large enough $r(tK_k) = (2k-1)t + r(K_k) - 2$ and $r(tC_k) = (2k - \lfloor \frac{k}{2} \rfloor)t - 1$. In [9] it is shown that $r(tC_4) = 6t - 1$ for all t.

2.2.2 General Graphs

We will give here some general upper bounds for the disjoint union of graphs and especially for multiple copies of the same graph. As an application we will determine $s(nK_t)$.

Lemma 2.21. *Let H be a graph containing at least one edge. Every H-minimal graph G has an edge-coloring with only red copies of H, and there are no two edge-disjoint red H.*

Proof. Let $G \to H$ minimal and $e \in E(G)$. Then $G - e \not\to H$ and therefore there exists an edge-coloring c of $G-e$ without monochromatic H. Let c' be the extension of c to G where e receives the color red. All monochromatic H have to contain e and thus there are no two edge-disjoint monochromatic H and all monochromatic H are red. $\square$

Corollary 2.22. *Let H be a graph containing at least one edge. Then H and $H + H$ are Ramsey-separable.*

Corollary 2.23. *Let H be a connected graph and $n \in \mathbb{N}$. Then*
$$s(nH) \leq s(H).$$

Proof. If H is an isolated vertex then the statement is trivial. Therefore we can assume that H contains at least one edge. Let G be a H-minimal graph and let c_{red} (c_{blue}) be the edge-coloring from Lemma 2.21 with only red (blue) copies of H, and there are no two edge-disjoint monochromatic H. Define $G' = (2n-1)G$. We will show that G' is nH-minimal.

In every edge-coloring of G' we have at least $2n-1$ disjoint monochromatic copies of H. By the pigeonhole principle there is a color, say red, such that at least n of these disjoint copies are completely red. This shows that $G' \to nH$.

We name the copies of G in G' by $G_1, G_2, \ldots, G_{2n-1}$. Let $e \in E(G')$, say $e \in E(G_1)$ without loss of generality. Then because G is H-minimal we can color $G_1 - e$ without monochromatic H. For $G_2, \ldots, G_n$ we take the coloring c_{red} and for $G_{n+1}, \ldots, G_{2n-1}$ we take the coloring c_{blue}. Since H is connected we can only take $n-1$ disjoint red H and only $n-1$ disjoint blue H which proves that $G' - e \not\to nH$.

Thus G' is nH-minimal and clearly $\delta(G') = \delta(G)$. □

Theorem 2.24. *For $t \geq 2$*
$$s(nK_t) = (t-1)^2.$$

Proof. By the above lemma and the fact that $s(K_t) = (t-1)^2$ it is enough to prove $s(nK_t) \geq s(K_t)$. Assume for contradiction that $s(nK_t) < (t-1)^2$. Then there exists a nK_t-minimal graph G and $x \in V(G)$ with $\deg(x) = \delta(G) < (t-1)^2$. Let c be any edge-coloring of $G - x$ without monochromatic copy of nK_t, i.e., there are at most $n-1$ red copies of K_t and at most $n-1$ blue copies of K_t. Let $R_1, \ldots, R_k$ be a maximal vertex-disjoint collection of red K_{t-1} in $N(x)$. Because $|N(x)| = \delta(G) < (t-1)^2$ we have $k < t-1$. We extend the coloring

2.2. Disjoint Union of Graphs

c to G by coloring every edge $\{x,y\}$ with $y \in \bigcup_i R_i$ blue and all the remaining edges red. This edge-coloring c' of G does not contain any new red K_t by the maximality and it does not contain any new blue K_t because $k < t-1$, which shows $G \not\to nK_t$. □

We continue by giving an upper bound for the disjoint union of two connected graphs G, H where we additionally assume that G is a supergraph of H. Theorem 2.25 is a generalization of Corollary 2.23 for $n = 2$, but not for $n > 2$ because of the condition about connectedness.

Theorem 2.25. *Let $G \supseteq H$ be two connected graphs. Then*
$$s(G+H) \leq \max\{s(G), s(H)\}.$$

Proof. Let K be a G-minimal graph with $\delta(K) = s(G)$ and J be a H-minimal graph with $\delta(J) = s(H)$. We proceed by giving a case distinction. Note that if we want to find a monochromatic copy of $G+H$ in a disjoint sum of graphs $G_1 + \ldots G_k$ then we have to find G in one G_i and H in one G_j because G, H are connected.

First, we assume that
$$J \to G. \tag{2.3}$$
Then clearly $\delta(J+J+J) = \delta(J) = s(H)$ and J is also a G-minimal graph because G contains H. Moreover, $J+J+J \to G+H$ because we find in each copy of J a monochromatic $G \supseteq H$ and at least two of them have the same color. For any edge e, we have $J-e \not\to H \subseteq G$. Let e be any edge of J. Then we can choose for one J a coloring without monochromatic $G+H$ and only containing red $H \subseteq G$ by Lemma 2.21. For the second J we can choose a coloring without monochromatic $G+H$ and only blue $H \subseteq G$ by Lemma 2.21. We can choose a coloring of $J-e$ without monochromatic H. This together is a coloring of $J+J+(J-e)$ without monochromatic $G+H$.

Thus, from now on, we can assume that $J \not\to G$, i.e., there is an edge-coloring of J without monochromatic G. Second, we assume that
$$K \to G+H. \tag{2.4}$$

Then K is also $(G+H)$-minimal and has minimum degree $\delta(K) = s(G)$.
Now, we assume that (2.4) and (2.5) are wrong and

$$K + J \to G + H. \qquad (2.5)$$

For $e \in E(K)$ we can color $K - e$ and J without monochromatic G showing that $(K - e) + J \not\to G \subseteq G + H$. For $e \in E(J)$ we can color K without monochromatic $G+H$ and $J-e$ without monochromatic H showing that $K + (J - e) \not\to G + H$.

In the next case, we assume that (2.4)-(2.6) are wrong and

$$K + J + J \to G + H. \qquad (2.6)$$

For $e \in E(J)$ we can color $J - e$ without monochromatic H and $K+J$ without monochromatic $G + H$ showing that $K + J + (J - e) \not\to G + H$. Next, assume that e is an edge from K. Because we assume that (2.4) is wrong we have that $J \not\to G$ and with $K - e \not\to G$ we have $(K - e) + J + J \not\to G \subseteq G + H$ showing that $K + J + J$ is $(G+H)$-minimal.

We continue by assuming that (2.4)-(2.7) are wrong and there is a subgraph K' of K (possibly K) such that

$$K + K' \to G + H. \qquad (2.7)$$

Let K' be a minimal such subgraph with respect to the number of edges. We have $\delta(K + K') \leq \delta(K) = s(G)$. For $e \in E(K')$, by the minimality $K + (K' - e) \not\to G + H$. If $K' = K$ then this is the only case to check. On the other hand K' is a proper subgraph of K and we have therefore $K' \not\to G$. This together with the fact that $K - e \not\to G$ implies that $(K - e) + K' \not\to G \subseteq G + H$.

Finally, we assume that (2.4)-(2.8) are wrong and there is a minimal subgraph K' of K such that

$$K + K' + J \to G + H, \qquad (2.8)$$

We have $\delta(K+K'+J) \leq \max\{\delta(K), \delta(J)\}$ and for $e \in E(K')$, by the minimality, $K+(K'-e)+J \not\to G+H$. For $e \in E(J)$ we have $J-e \not\to H$.

This together with the assumption $K + K' \not\to G + H$ implies that there is a coloring of $K + K' + (J - e)$ without monochromatic $G + H$. If $K' = K$ then these are the only two cases to check. Otherwise K' is a proper subgraph of K and therefore $K' \not\to G$ as well as $K - e \not\to G$. This together with the assumption that $J \not\to G$ implies that $(K - e) + K' + J \not\to G \subseteq G + H$.

One of the above cases apply because we always have that $K + K + J \to G + H$: There is a monochromatic copy of G in both copies of K and if they both have the same color then we are done because $H \subseteq G$. Otherwise we have a red G and a blue G which with the monochromatic copy of H in J completes the proof. □

If $G \supseteq H$ are two connected graphs such that G and H are Ramsey-separable, then we have $s(G + H) \leq s(G)$. The reason is that then we can choose J such that (2.4) is not true and proceed with the proof as before.

2.3 No upper bound for s in terms of δ

In Section 2.1 we have seen a large class of bipartite graphs G with $s(G) = 2\delta(G) - 1$ and moreover it is known that for cliques $s(K_k) = \delta(K_k)^2$. The question arises whether there is a function f such that for all graphs G it holds that $s(G) \leq f(\delta(G))$. Actually, we have already seen that such a function cannot exist. Namely, for $G = K_t + K_2, t \geq 4$ we have $\delta(G) = 1$ and $s(G)$ arbitrarily large (Theorem 2.16). This family contains only non-bipartite graphs which are disconnected. Are these conditions somehow necessary? One could try to take some bipartite graph H in the disjoint union with K_2. This attempt fails:

Proposition 2.26. *Let H be a bipartite graph with $\delta(H) \geq 1$. Then $s(H + K_2) = 1$.*

Proof. Let (A, B) be a bipartition of H such that $|B| = a(H)$ and let x, y be the vertices of the K_2. Then $(A \cup \{x\}, B \cup \{y\})$ is a bipartition of

$H + K_2$ and $a(H + K_2) = a(H) + 1 = |B| + 1$. Applying Theorem 2.10 finishes the proof. □

We do not know whether there exists a bipartite graph not attaining the simple lower bound (Proposition 2.2), but we will show in this section that there exist *connected* graphs with minimum degree 1 and the s-value arbitrarily large.

Let $K_t \cdot K_2$ be the graph containing a t-clique and one additional vertex connected to exactly one vertex of the t-clique. We call this additional edge and additional vertex a *hanging edge* and a *hanging vertex*, respectively. Clearly, $K_t \cdot K_2$ is connected and has minimum degree 1.

Let us assume that we are given a graph H with some red/blue edge-coloring c without a monochromatic $K_t \cdot K_2$ and assume from now on that $t \geq 3$. We call a vertex which is incident to two monochromatic copies of K_t *critical*, a vertex which is incident to one monochromatic copy of K_t *harmless*, and other vertices *safe*. For a vertex $u \in V(H)$, we denote the set of those neighbors of u which are adjacent to u via a red (blue) edge by $R(u)$ $(B(u))$.

Lemma 2.27. *Let $t \geq 3$ and c a red/blue edge-coloring without monochromatic $K_t \cdot K_2$ and $u \in V(H)$ be a critical vertex. Then*

(i) $|R(u)| = |B(u)| = t - 1$, *i.e., u is contained in exactly one red and exactly one blue copy of K_t and has no other incident edges.*

(ii) $E(R(u), B(u)) = \emptyset$, *i.e., there is no edge between the red neighbors of u and the blue neighbors of u.*

Proof. (i) Since u is critical it has to be incident to two monochromatic K_t. If they were in the same color then this would yield a monochromatic $K_t \cdot K_2$. Moreover, there cannot be more edges incident to u without creating a monochromatic $K_t \cdot K_2$.

(ii) Assume there is an edge between a red neighbor of u and a blue neighbor of u. Then this edge has some color according to c and

2.3. No upper bound for s in terms of δ

therefore it completes a monochromatic $K_t \cdot K_2$ in this color with one of the t-cliques containing u. □

Lemma 2.28. *For every graph H and every edge-coloring c without monochromatic $K_t \cdot K_2$ there exists a new edge-coloring c_F with no critical vertices and no monochromatic $K_t \cdot K_2$.*

Proof. For each t-clique which is monochromatic in c and has a critical vertex, choose one arbitrary edge containing a critical vertex. Let us denote the set of these edges by $F \subseteq E(H)$ and let c_F be the coloring obtained from c after we change the color of each edge in F.

By Lemma 2.27, we see that every critical vertex u is incident to exactly two t-cliques and none of them is monochromatic in the coloring c_F, since we changed the color of exactly one edge in each. Thus each critical vertex in c is safe in c_F.

Every edge whose color was changed contains a critical vertex in c, which is safe in c_F, meaning that these edges cannot be part of a new monochromatic K_t. That is, every monochromatic K_t in c_F was already monochromatic in c and hence no vertices are critical in c_F.

We still need to show that there is no monochromatic $K_t \cdot K_2$ in c_F. Since every monochromatic K_t in c_F was monochromatic in c, and c has no monochromatic $K_t \cdot K_2$, the only possibility to have a monochromatic $K_t \cdot K_2$ in c_F would be that the hanging edge e changed its color. Let U be the vertex set of the K_t within a monochromatic, say blue, $K_t \cdot K_2$ in c_F. Then the edges within U were already blue in c, while e was red in c. Since e changed its color, it was part of a red t-clique W in c. Hence $U \cap W = \{u\}$, where u is an endpoint of e, and u is critical in c. This is a contradiction, since u is not safe in c_F. □

Theorem 2.29. *For every $t \geq 3$,*

$$s(K_t \cdot K_2) \geq t - 1 \, .$$

Proof. Suppose for contradiction that there is a $K_t \cdot K_2$-minimal graph G with a vertex x of degree less than $t - 1$. Since $G - x$ is not $K_t \cdot$

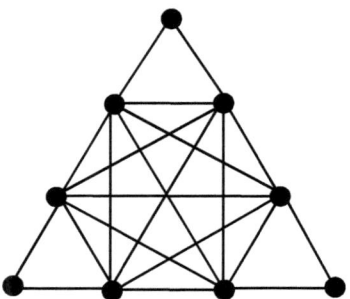

Figure 2.2: $s(K_3 \cdot K_2) \leq 2$

K_2-Ramsey, there is a red/blue edge-coloring c of $G - x$ without a monochromatic $K_t \cdot K_2$. By Lemma 2.28 we can also assume that c has no critical vertices. We now extend c to a coloring of G. Color each edge $\{x, y\}$ of G red if y is contained in a blue t-clique of c, and blue otherwise. Since there are no critical vertices, y cannot be contained in a blue *and* a red t-clique, therefore the coloring extension is well-defined.

Since x has degree less than $t - 1$, it can contribute to a monochromatic $K_t \cdot K_2$ only as a hanging vertex. Let e be the hanging edge of a monochromatic $K_t \cdot K_2$ containing x and let U be the monochromatic t-clique. By the definition of the extended coloring, the colors of the edge e and of the edges in U are different, a contradiction. □

The lower bound is tight for $t = 3$:

Proposition 2.30.
$$s(K_3 \cdot K_2) = 2.$$

Proof. By Theorem 2.29 we know that $s(K_3 \cdot K_2) \geq 2$. Therefore it is enough to give a $K_3 \cdot K_2$-minimal graph G with minimum degree 2.

We extend K_6 by three paths of length 2, see Figure 2.2, and claim that this graph G is $K_3 \cdot K_2$-minimal. In any red/blue edge-coloring of K_6 there is a monochromatic triangle. It is possible to color K_6 without

2.4. Cliques—More Colors

a monochromatic $K_3 \cdot K_2$, namely coloring two disjoint triangles blue and coloring all other edges red. It is easy to see that, up to renaming of the vertices and the colors, this is the only such edge-coloring. In any partition of the K_6 of G into two triangles T_1, T_2, there is a path P of length 2 connecting a vertex in T_1 and a vertex in T_2. Suppose that the edges of T_1 and T_2 are colored blue. If one of the edges of the path P is also blue, then this would complete a blue $K_3 \cdot K_2$. Thus both edges of the path P are colored red. Hence P and the other red edges going between T_1 and T_2 yield a red $K_3 \cdot K_2$. This shows that the graph G is $K_3 \cdot K_2$-Ramsey.

If we delete an edge from the K_6 then we can color K_6 without monochromatic triangles and extend this to each of the three 2-paths by coloring their two edges with distinct colors. If we delete an edge e lying on a 2-path of G then it is easy to color $G - e$ without creating a monochromatic $K_3 \cdot K_2$: partition the vertex set of the K_6 of G into two blue triangles such that each of the endpoints of the remaining two 2-paths are completely contained in one of the blue triangles and color all other edges red. It is easy to see that this edge-coloring has no monochromatic $K_3 \cdot K_2$. Hence G is $(K_3 \cdot K_2)$-minimal. $\square$

Actually, we proved a stronger statement, namely the graph G shown in Figure 2.2 is $K_3 \cdot K_2$-minimal.

2.4 Cliques—More Colors

As it is quite usual in Ramsey theory, one can consider generalizations of results with more than two colors and asymmetric cases. The results for bipartite graphs can be generalized to asymmetric cases as well as to more than two colors. Because of the lack of new insights we leave out the details here. Instead, we will consider cliques in more than two colors after we introduce the necessary definitions. While we can give a general upper bound for the s-value of the tuple $(K_{a_1}, \ldots, K_{a_r})$, its tightness is proven only for special cases.

Definition 2.31. Let $r \in \mathbb{N}$ and $G, H_1, H_2, \ldots, H_r$ be graphs. If in every edge-r-coloring of G with colors $1, 2, \ldots, r$ there is $j \in [r]$ and a copy of H_j completely in color j, then we say that G is $(H_1, H_2, \ldots, H_r)$-Ramsey and write $G \to (H_1, H_2, \ldots, H_r)$. If additionally every proper subgraph of G is not $(H_1, H_2, \ldots, H_r)$-Ramsey, then we say that G is $(H_1, H_2, \ldots, H_r)$-minimal. Define

$$s(H_1, \ldots, H_r) = \min\{\delta(G) : G \text{ is } (H_1, \ldots, H_r)\text{-minimal}\}.$$

If we have only three colors then we assume that these three colors are named red, blue, green. The value $s(K_{a_1}, \ldots, K_{a_r})$ was previously only known for $r = 2$, i.e., $s(K_a, K_b) = (a-1)(b-1)$, see [19, 40].

2.4.1 Upper Bound

The upper bound shown here is a generalization of the upper bound construction for two colors in the symmetric setting given by Fox and Lin [40].

Let G be a graph and $k, r \in \mathbb{N}$. Let $\mathcal{F}(G, k, r)$ be the family of graphs F that satisfy (i) $\omega(F) = \omega(G)$, and (ii) in every vertex-coloring of F with k colors and in every edge-coloring of F with r colors there exists a copy of G that is monochromatic in the edge-coloring and it is monochromatic in the vertex-coloring.

Folkman [39] proved that $\mathcal{F}(G, k, 1)$ is non-empty for every graph G and $\mathcal{F}(K_s, 1, 2)$ is non-empty for every $s \in \mathbb{N}$. He conjectured that the second result should also be true for more than two colors, which was proven by Nešetřil and Rödl [70]. Moreover, they proved $\mathcal{F}(G, 1, r) \neq \emptyset$ for every graph G. It is easy to see now that the family $\mathcal{F}(G, k, r)$ always contains at least one graph: Let $G_1 \in \mathcal{F}(G, k, 1)$ and $G_2 \in \mathcal{F}(G_1, 1, r)$ then $G_2 \in \mathcal{F}(G, k, r)$.

The second main tool is the lexicographic product, sometimes also called the Abbott product because of its application in [2] (do not google the term "Abbott product").

2.4. Cliques—More Colors

Definition 2.32. *Let A, B be graphs. The* lexicographic product $A \ltimes B$ *is the graph whose vertex set is $V(A) \times V(B)$, with edges given by (a, b) is adjacent to (a', b') if either a is adjacent to a' in A or $a = a'$ and b is adjacent to b' in B.*

Let $G = A \ltimes B$ be the lexicographic product of the graphs A and B. For $x \in V(A)$ we define $B_x = G[S]$ to be the induced subgraph of G on the vertices $S = \{x\} \times V(B)$. It is obvious that $B_x \cong B$.

An alternative definition for the lexicographic product $A \ltimes B$ is the following: Let B_x be disjoint copies of B for $x \in V(A)$. Then for each pair $x, y \in V(A), \{x, y\} \in E(A)$ add all the edges between $V(B_x)$ and $V(B_y)$. We note that in general $A \ltimes B \not\cong B \ltimes A$ (Figure 2.3 indicates an example for that) and it is easy to see that $\omega(A \ltimes B) = \omega(A) \cdot \omega(B)$.

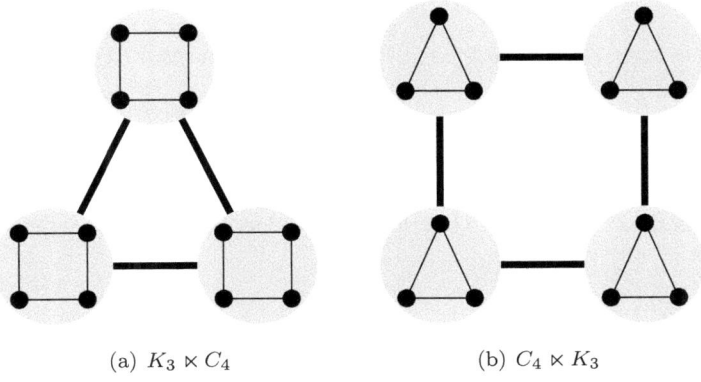

(a) $K_3 \ltimes C_4$ \qquad (b) $C_4 \ltimes K_3$

Figure 2.3: The lexicographic products $K_3 \ltimes C_4, C_4 \ltimes K_3$, where the fat lines indicated that all edges between these two parts are present.

Proposition 2.33. *The lexicograhic product is associative, i.e., for graphs A, B, C we have $A \ltimes (B \ltimes C) = (A \ltimes B) \ltimes C$.*

Proof. It is clear that the vertex set of both graphs is $V(A) \times V(B) \times V(C)$. A pair of vertices $(a, b, c), (a', b', c')$ is adjacent for both graphs if either (i) $\{a, a'\} \in E(A)$, or (ii) $a = a'$ and $\{b, b'\} \in E(B)$, or (iii) $a = a', b = b'$ and $\{c, c'\} \in E(C)$. This shows that the graphs are the same. □

Since the lexicographic product is associative, we will leave out the round brackets in the following and just write $A \ltimes B \ltimes C$.

Definition 2.34. *Let A, B, C be graphs. An edge-r-coloring φ of $A \ltimes B$ is good with respect to C if for all $x \in V(A)$ there exists $V_x \subseteq V(B_x)$ such that*

(i) $C_x := B_x[V_x] \cong C$ for all $x \in V(A)$, and
(ii) for all $\{a_1, a_2\} \in E(A)$ all edges between V_{a_1} and V_{a_2} have the same color.

We write

$$A \ltimes B \to^{r\text{-}good} A \ltimes C,$$

if every edge-r-coloring of $A \ltimes B$ is good with respect to C.

Note that for an edge-coloring of $A \ltimes B$ to be good with respect to C, it does not matter how the edges inside the B_x are colored. If φ is a good edge-r-coloring of $A \ltimes B$ as above then there is an edge-r-coloring φ' of A such that for every edge $\{a_1, a_2\} \in E(A)$ we have $\varphi(E(V_{a_1}, V_{a_2})) = \{\varphi'(\{a_1, a_2\})\}$.

Lemma 2.35. *Let A, C be graphs and let $r \geq 1$ be an integer. Then there exists B with $\omega(B) = \omega(C)$ such that*

$$A \ltimes B \to^{r\text{-}good} A \ltimes C.$$

Proof. Let $B' \in \mathcal{F}(C, d_1, 1), B \in \mathcal{F}(B', d_2, 1)$ for integers d_1, d_2 which will be defined later. By definition of $\mathcal{F}$ we have $\omega(B) = \omega(C)$.

Let c be an edge-coloring of $A \ltimes B$ with r colors. We want to show that c is good with respect to C. Without loss of generality we can assume that $V(A) = \{1, 2, \ldots, n\}$. As introduced before we will use the notation B_j for the induced subgraph of $A \ltimes B$ on vertex set $\{j\} \times V(B)$. We will proceed in two steps.

2.4. Cliques—More Colors

Claim 1. *There are $W_1 \subseteq V(B_1), \ldots, W_n \subseteq V(B_n)$ such that*

(i) *for all j, $B_j[W_j] \cong B'$, and*
(ii) *for $2 \leq j \leq n$ and $x \in \bigcup_{k<j} W_k$ all edges between x and W_j have the same color.*

Proof. We begin with any $W_1 \subseteq V(B_1)$ with $B_1[W_1] \cong B'$ which exists because $B_1 \in \mathcal{F}(B', d_2, 1)$. We will continue inductively and therefore we can assume that for $1 < m < n$ we have found $W_1, \ldots, W_m$ satisfying (i) and (ii). Let $S = \bigcup_{1 \leq j \leq m} W_j$ and assume for technical reasons that S is ordered. We associate with each vertex $x \in V(B_{m+1})$ a vector $(c(\{x,y\}))_{y \in S} \in [r]^{|S|}$ which contains the colors of its incident edges (we set $c(\{x,y\})$ to some arbitrary color if $\{x,y\}$ is not an edge). Define a vertex-coloring φ of B_{m+1} by $\varphi(x) = (c(\{x,y\}))_{y \in S}$ with $r^{|S|} = r^{mn(B')}$ colors. Because $B_{m+1} \in \mathcal{F}(B', d_2, 1)$ and we can choose $d_2 = r^{(n(A)-1)n(B')}$ there is a vertex set $W_{m+1} \subseteq B_{m+1}$ such that $B_{m+1}[W_{m+1}] \cong B'$ and it is monochromatic under φ. Clearly, (i) holds now for $j = m+1$. Since all $y \in W_{m+1}$ have the same color under φ, condition (ii) follows for $j = m+1$ as well. Denote by B'_j the induced subgraph of B_j on the vertex set W_j. This finishes the proof of the claim. □

We can repeat this reasoning in the backward order. Since the proof is exactly the same, we will only statement.

Claim 2. *There are $V_n \subseteq W_n, \ldots, V_1 \subseteq W_1$ such that*

(i') *for all j, $C_j := B'_j[V_j] \cong C$, and*
(ii') *for $1 \leq j \leq n-1$ and $x \in \bigcup_{k>j} V_k$ all edges between x and V_j have the same color.*

Note that this time we can choose $d_1 = r^{(n(A)-1)n(C)}$. The conditions (ii) and (ii') imply that for every pair $\{a_1, a_2\} \in E(A)$ the edges between V_{a_1} and V_{a_2} have the same color. □

Definition 2.36. Let $A_1, A_2, \ldots, A_s$ be graphs. Let G be the lexicographic product $A_1 \ltimes A_2 \ltimes \ldots \ltimes A_s$ and define for $1 \leq i \leq s$

$$E_i = \{\{x,y\} \in E(G) : x_i \neq y_i, x_j = y_j \forall j < i\},$$

and for $t_1 \in V(A_1), \ldots, t_{i-1} \in V(A_{i-1})$

$$E_i(t_1, \ldots, t_{i-1}) = \{\{x,y\} \in E(G) : x_i \neq y_i, x_j = y_j = t_j \forall j < i\}.$$

An edge-r-coloring c of G is 1-uniform if there exists an edge-r-coloring c_1 of A_1 such that $c(\{x,y\}) = c_1(\{x_j, y_j\})$ for all $\{x,y\} \in E_1$; and for $i \geq 2$, an edge-r-coloring c of G is i-uniform if for all $t_1, \ldots, t_{i-1}$ there exists an edge-r-coloring $c_{i,t_1,\ldots,t_{i-1}}$ of A_i such that $c(\{x,y\}) = c_{i,t_1,\ldots,t_{i-1}}(\{x_j, y_j\})$ for all $\{x,y\} \in E_i(t_1, \ldots, t_{i-1})$. Moreover, if c is i-uniform for all $1 \leq i \leq s$, then we call c totally uniform.

For example, if we color *all* edges by the same color then this is a totally uniform coloring. Another example is to color all the edges in E_i by color i. But there is more freedom for a totally uniform coloring as, for example, Figure 2.4 shows. Note that every coloring is s-uniform. The notion of totally uniform coloring is a generalization of good colorings for the lexicographic product of more than two graphs. For $s = 2$, an edge-r-coloring of $A_1 \ltimes A_2$ is good with respect to A_2 if it is totally uniform. There is a generalization of Lemma 2.35.

Lemma 2.37. Let $r \geq 1, s \geq 2$ and $A_1, A_2, \ldots, A_s$ be graphs. There are graphs $B_2, \ldots, B_s$ with $\omega(B_i) = \omega(A_i)$ for $i = 2, \ldots, s$ such that for every edge-r-coloring c of $A_1 \ltimes B_2 \ltimes \ldots \ltimes B_s$ there are subgraphs A'_i of B_i where $i = 2, \ldots s$ with $A'_i \cong A_i$ and c is totally uniform on $A_1 \ltimes A'_2 \ltimes \ldots \ltimes A'_s$.

Proof. We proceed by an induction over s. For $s = 2$ the statement reduces to Lemma 2.35. Let $s > 2$. Assume as induction hypothesis that the statement is true for $s - 1$, i.e., there are $B_2, \ldots B_{s-1}$ such that $\omega(B_i) = \omega(A_i)$ for $i = 2, \ldots, s - 1$ and for every edge-r-coloring of

2.4. Cliques—More Colors

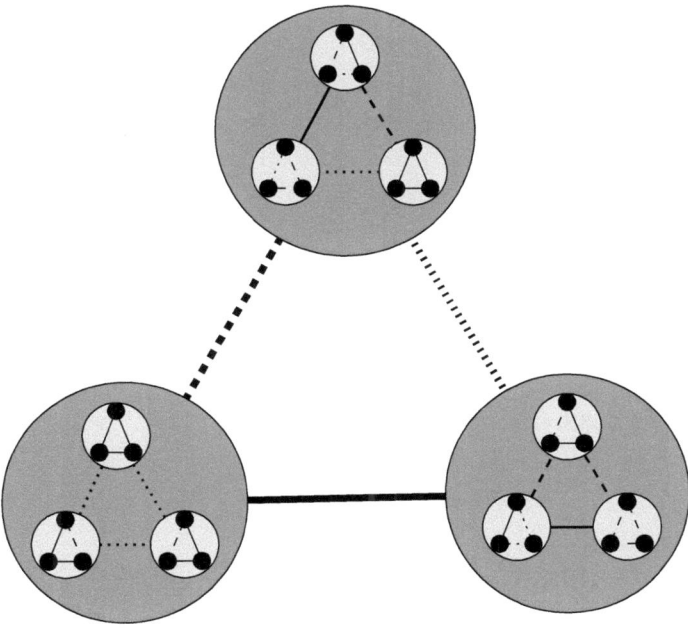

Figure 2.4: A totally uniform coloring of $K_3 \ltimes K_3 \ltimes K_3$. Different colors are indicated by solid, dashed, and dotted lines and thick lines represent that all edges between these two parts are colored the same.

$A_1 \ltimes B_1 \ltimes \ldots \ltimes B_{s-1}$ there are subgraphs A'_i of B_i with $A'_i \cong A_i$ such that c is totally uniform on $A_1 \ltimes A'_2 \ltimes \ldots \ltimes A'_{s-1}$.

By Lemma 2.35 we can choose B_s with $\omega(B_s) = \omega(A_s)$ such that

$$(A_1 \ltimes B_1 \ltimes \ldots \ltimes B_{s-1}) \ltimes B_s \to^{r\text{-good}} (A_1 \ltimes B_1 \ltimes \ldots \ltimes B_{s-1}) \ltimes A_s.$$

We claim that these $B_2, \ldots, B_{s-1}$ together with B_s fulfill the requirements of the lemma. Clearly, $\omega(B_i) = \omega(A_i)$ for $i = 2, \ldots, s$. Let c be an edge-r-coloring of $A_1 \ltimes B_2 \ltimes \ldots \ltimes B_s$. By the choice of B_s there is a subgraph A'_s of B_s with $A'_s \cong A_s$ such that c is good with respect to A_s. Thus, there is an edge-r-coloring c_s on $H' = A_1 \ltimes B_1 \ltimes \ldots \ltimes B_{s-1}$ such that for each edge $\{a_1, a_2\} \in E(H')$ we have $c(E((A'_s)_x, (A'_s)_y)) = \{c_s(\{x, y\})\}$. This shows that c is $(s-1)$-uniform on this graph. By induction hypothesis we know that there are subgraphs A'_i of B_i for $i = 2, \ldots, r-1$ such that c_s is totally uniform on $A_1 \ltimes A'_2 \ltimes \ldots \ltimes A'_{s-1}$. Therefore, c is totally uniform on $A_1 \ltimes A'_2 \ltimes \ldots \ltimes A'_s$. □

Definition 2.38. *An edge-r-coloring φ of $A_1 \ltimes A_2 \ltimes \ldots \ltimes A_s$ is totally uniformly monochromatic if all the edges in E_1 have the same color and for $2 \leq j \leq r; t_1 \in V(A_1), \ldots, t_{j-1} \in V(A_{j-1})$ all the edges in $E_j(t_1, \ldots, t_{j-1})$ have the same color.*

Corollary 2.39. *Let $r \geq 1, s \geq 2$ and $A_1, A_2, \ldots, A_s$ be graphs. There are graphs $C_1, C_2, \ldots, C_s$ with $\omega(C_i) = \omega(A_i)$ for $i = 1, \ldots, s$ such that for every edge-r-coloring φ of $H = C_1 \ltimes \ldots \ltimes C_r$ there are subgraphs A'_i of C_i with $A'_i \cong A_i$ such that φ is totally uniformly monochromatic on $A'_1 \ltimes \ldots \ltimes A'_s$.*

Proof. For $1 \leq i \leq s$ let $G_i \in \mathcal{F}(A_i, 1, r)$ and $C_1 = G_1$. For $2 \leq i \leq s$ choose C_i according to Lemma 2.37 with $\omega(C_i) = \omega(G_i) = \omega(A_i)$ such that in every edge-r-coloring c of $C_1 \ltimes C_2 \ltimes \ldots \ltimes C_s$ there are subgraphs G'_i of C_i with $G'_i \cong G_i$ such that c is totally uniform on $G_1 \ltimes G'_2 \ltimes \ldots \ltimes G'_s$. Since $G'_i \cong G_i \in \mathcal{F}(A_i, 1, r)$ the statement follows. □

2.4. Cliques—More Colors

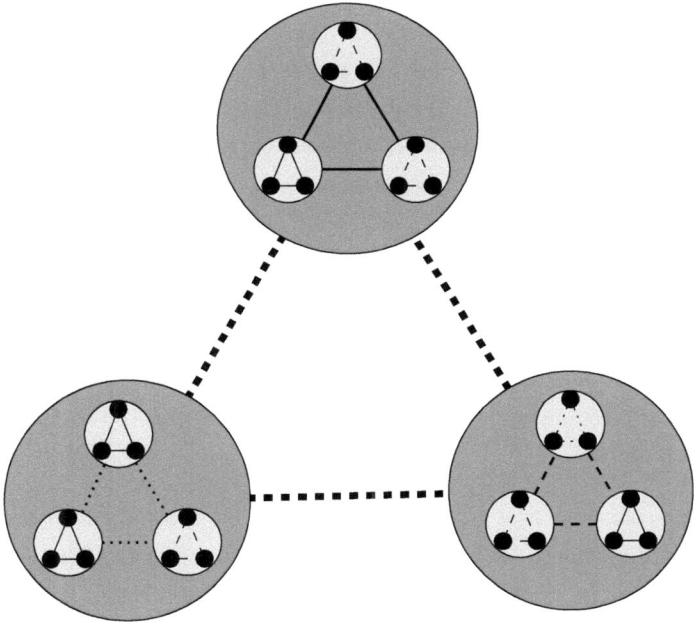

Figure 2.5: A totally uniformly monochromatic coloring of $K_3 \ltimes K_3 \ltimes K_3$. Different colors are indicated by solid, dashed, and dotted lines and thick lines represent that all edges between these two parts are colored the same.

Theorem 2.40. *Let $r \geq 1$. For $a_1 \geq \ldots \geq a_r \geq 1$*

$$s(K_{a_1}, \ldots, K_{a_r}) \leq \prod_{i=1}^{r}(a_i - 1) =: p.$$

Proof. To prove the theorem we will construct a $(K_{a_1}, \ldots, K_{a_r})$-minimal graph with minimum degree p.

Let $H = C_1 \ltimes C_2 \ltimes \ldots \ltimes C_r$ be the graph from Corollary 2.39 for $A_i = K_{a_i-1}$, $i = 1, \ldots, r$. First, we prove $H \not\to (K_{a_1}, \ldots, K_{a_r})$. Look at the coloring of H where E_i is colored completely in color i. The ith color class in this coloring contains the edges corresponding to a subgraph isomorphic to $mC_i \ltimes nK_1$ for some $m, n \in \mathbb{N}$. There is no monochromatic a_i-clique in color i because $\omega(mC_i \ltimes nK_1) = \omega(C_i) = a_i - 1$.

Let $H_1 = \mathcal{T}_p^1(H; V(H))$, i.e., we add for each p-tuple of $V(H)$ a new vertex and connect it to the members of the tuple. For every edge-r-coloring c of H there are subgraphs A_i of C_i with $A_i \cong K_{a_i-1}$ such that c is totally uniformly monochromatic on $A_1 \ltimes \ldots \ltimes A_r$. Moreover, there is a newly introduced vertex w in H_1 which is connected to all vertices of that subgraph.

Claim 3. *Assume that $G = K_{a_1-1} \ltimes \ldots \ltimes K_{a_r-1}$ is totally uniformly monochromatic colored with r colors and w is connected to all its vertices. Every r-coloring of the edges incident to w completes the edge-r-coloring such that there is for some i with $1 \leq i \leq r$ an a_i-clique completely in color i.*

Proof. We proceed by induction over r. The statement is trivial for $r = 1$. Let $r > 1$ and assume that the statement holds for $r - 1$. All the edges in E_1 of G have the same color, say color j. If $a_j < a_1$ then there is already an a_j-clique completely in color j. On the other hand we have $a_j = a_1$, because the a_i's are ordered, and we can assume that by renaming the colors $j = 1$ holds. If for some x there is an edge in color 1 inside $(K_{a_2-1} \ltimes \ldots \ltimes K_{a_r-1})_x$ then there is a monochromatic a_1-clique in color 1 and we are done. Thus, we can assume there is no edge in

2.4. Cliques—More Colors

color 1 except the ones in E_1. Furthermore, if w is connected to every $(K_{a_2-1} \ltimes \ldots \ltimes K_{a_r-1})_x$ by at least one edge in color 1, then this finishes a monochromatic K_{a_1} completely in color 1. Thus, we can assume that there is an x such that all edges among w and $(K_{a_2-1} \ltimes \ldots \ltimes K_{a_r-1})_x$ are only using the colors $2, \ldots, r$. By induction we know that there is an a_j-clique completely in color j for $2 \leq j \leq r$. □

Therefore, $H_1 \to (K_{a_1}, \ldots, K_{a_2})$. There exists a Ramsey-minimal graph H_2 with $H_2 \subseteq H_1$. Not all of the newly introduced vertices in H_1 are deleted in H_2, because otherwise $H_2 \subseteq H \not\to (K_{a_1}, \ldots, K_{a_2})$. Thus, the minimum degree of H_2 is at most p. □

2.4.2 Lower Bounds

For providing lower bounds we try to adapt the technique from Theorem 2.24 and thereby prove stronger statements about some special colorings of cliques.

Definition 2.41. *Let $H, G_1, \ldots, G_r$ be graphs and let c be an edge-r-coloring of H. Then we say that c is a $(G_1, \ldots, G_r)$-good coloring if there is an vertex-r-coloring of H, such that there is no copy of G_j with edges and vertices of color j only, for every $1 \leq j \leq r$. If c is not a $(G_1, \ldots, G_r)$-good coloring then we say that it is a $(G_1, \ldots, G_r)$-critical coloring. For $G_j = K_{a_j}$ we just write $(a_1, \ldots, a_r)$-good and $(a_1, \ldots, a_r)$-critical, respectively.*

If c is a $(G_1, \ldots, G_r)$-critical coloring of H with $|V(H)| = k$, then there exists also a critical coloring of K_k by arbitrarily extending c. Therefore we will only look at complete graphs and their colorings.

Definition 2.42. *Let $s^*(G_1, \ldots, G_r)$ be the minimum k such that there exists a $(G_1, \ldots, G_r)$-critical edge-coloring of K_k.*

Let us note that such an edge-coloring c can contain large monochromatic cliques, but we want large cliques that are monochromatic in the vertex- and edge-coloring.

Proposition 2.43.

(i) $\quad s^*(1, a_2, \ldots, a_r) = s^*(a_2, \ldots, a_r).$

(ii) $\quad s^*(a_1 - 1, \ldots, a_r - 1) \leq s(a_1, \ldots, a_r).$

(iii) $\quad s^*(a_1, a_2, \ldots, a_r) \geq (\sum_{i=2}^{r} a_i - r + 2)a_1.$

(iv) $\quad s^*(a, b) = ab.$

(v) $\quad s^*(a_1, \ldots, a_r) \leq \prod_{i=1}^{r} a_i.$

Proof. (i) An $(1, a_2, \ldots, a_r)$-critical coloring is also $(a_2, \ldots, a_r)$-critical and vice versa.

(ii) Let G be a $(a_1, \ldots, a_r)$-minimal graph with minimum degree $\delta(G) = s(a_1, \ldots, a_r)$ and $x \in V(G)$ with $\deg(x) = \delta(G)$. Let $H = G[N(x)]$ and c an edge-r-coloring of $G - x$ without monochromatic a_j-clique in color j for all j. The restriction of c to the edges of H is $(a_1 - 1, \ldots, a_r - 1)$-critical, because any vertex-k-coloring of H can be viewed as a r-coloring of the edges incident to x in G. We have $n(H) = \delta(G)$ and as above we can extend this edge-coloring of H to a complete graph with $n(H)$ vertices which is still critical. The desired inequality follows.

(iii) Let c be any edge-coloring of K_n with $n < (\sum_{i=2}^{r}(a_i - 1) + 1)a_1$. Let $R_1, \ldots, R_k$ be a maximal vertex disjoint collection of a_1-cliques which are monochromatic in the first color. We have $k \leq \sum_{i=2}^{r}(a_i - 1)$. Define $s_1 = 0, s_m = \sum_{i=2}^{m}(a_i - 1)$. Color the vertices of $\bigcup_{j=s_{m-1}+1}^{s_m} R_j$ by color m and the remaining vertices by color 1. There is no a_1-clique with vertices and edges in color 1 by the maximality and there are at most $(a_m - 1)$-cliques in color m.

(iv) By (iii) it follows that $s^*(a, b) \geq ab$. Together with (i) and the result $s(a+1, b+1) = ab$ this implies the statement.

(v) It is simple to check that the lexicographical coloring of the graph $K_{a_1}[K_{a_2}] \ldots [K_{a_r}]$ is $(a_1, a_2, \ldots, a_r)$-critical. □

2.4. Cliques—More Colors

For $a_1 = a_2 = \ldots = a_r = 2$, we have

$$s^*(2, 2, \ldots, 2) \leq s(3, 3, \ldots 3) \leq 2^r.$$

Also we suspect that 2^r is the right value for the s-parameter, we can show that for the s^* it is actually much smaller.

Proposition 2.44. *For $r \geq 1$ it holds*

$$s^*(\underbrace{2, 2, \ldots, 2}_{r}) \leq 2r^2 \log r + 1.$$

Proof. We construct a $(2, 2, \ldots, 2)$-critical edge-coloring randomly. Let $n = 2r^2 \log r + 1$ and look at a random edge-r-coloring φ of K_n (choose the color of each edge uniformly at random). Then for a fixed vertex-r-coloring γ of K_n denote the event that there is no monochromatic edge in φ with its endpoints of the same color in γ by M_γ. By using Jensen's inequality, we have

$$\Pr[M_\gamma] = \prod_{j=1}^{r} \Pr[\gamma^{-1}(j) \text{ has no edge in color j under } \varphi]$$

$$= \prod_{j=1}^{r} (1 - \frac{1}{r})^{\binom{|\gamma^{-1}(j)|}{2}} = (1 - \frac{1}{r})^{\sum_{j=1}^{r} \binom{|\gamma^{-1}(j)|}{2}}$$

$$\leq (1 - \frac{1}{r})^{r \binom{\frac{1}{r} \sum_{j=1}^{r} |\gamma^{-1}(j)|}{2}}$$

$$< \exp\left(-\binom{n/r}{2}\right).$$

Thus the expected number of monochromatic edges under any γ is

$$\sum_\gamma \Pr[M_\gamma] < r^n \exp\left(-\binom{n/r}{2}\right)$$

$$= \exp\left(n \log r - \binom{n/r}{2}\right) \leq 1.$$

This random edge-r-coloring of K_n has less than 1 monochromatic edge in expectation. Since the number of monochromatic edges is for every fixed φ a nonnegative integer, there has to be at least one edge-r-coloring φ with no monochromatic edge under any vertex-r-coloring γ. □

Lemma 2.45. *The only graph G on at most 4 vertices with a $(2,2)$-critical coloring is K_4.*

Proof. It is enough to prove that every edge-coloring of $G = K_4 - e$ is $(2,2)$-good. Let a,b be non-adjacent vertices in G and let c,d be the other two vertices. For an edge-coloring χ of G with $\chi(\{c,d\}) = $ red, color the vertices c,d blue and color a,b red. For an edge-coloring χ of G with $\chi(\{c,d\}) = $ blue, color c,d red and a,b blue. In both colorings there is no monochromatic edge with its endpoints in the same color, which proves that all edge-colorings of G are $(2,2)$-good. □

Proposition 2.46. $s^*(2,2,2) \geq 8$.

Proof. We show that every edge-3-coloring of K_7 is $(2,2,2)$-good. The monochromatic degree of a vertex $x \in V(G)$ is at least 2 for some color, say green. Let y,z be the neighbors of x that are connected by a green edge and assume that $\{y,z\}$ is blue or green again. If there is a red edge in $G - \{x,y,z\}$ then by Lemma 2.45 we can color $V(G) \setminus \{x,y,z\}$ by using only blue and green and not create a monochromatic K_2. Then we can finish by coloring x,y,z red. On the other hand if there is no red edge in $G - \{x,y,z\}$ then we color $V(G) \setminus \{x,y,z\}$ red and x,y blue and y green. □

Lemma 2.47. *In every red/blue/green edge-coloring of K_5 which is not completely red, there exists a green or blue edge such that the remaining triangle is not polychromatic, i.e., it has at most two different colors.*

Proof. Assume for contradiction that there exists a red/blue/green edge-coloring of K_5 such that (i) there exists a blue or green edge, and (ii) for every blue/green edge the remaining 3 vertices form a polychromatic triangle.

First, we assume that there is a blue/green triangle T in this coloring. The remaining edge e in $K_5 - T$ has color red otherwise it would contradict (ii). For every edge in T the other 3 vertices have to form a polychromatic triangle by (ii). Thus all the edges between T and e

2.4. Cliques—More Colors 55

have to be blue or green. The edge connecting one endpoint of e and a vertex from T is blue or green but the remaining triangle has no red edge contradicting (ii).

Second, we assume that there is a red triangle T in this coloring. The remaining edge e has to be red as well by (ii). There has to be at least one blue or green edge by (i). It is easy to see now that *all* edges between T and e have to be blue or green. Let a, b be the endpoints of e. Without loss of generality the blue degree of a is at least two. These two neighbors together with a form a red/blue triangle and the remaining edge is blue or green, contradicting (ii).

Finally, we assume that there is neither a blue/green triangle nor a red triangle. Then there is a blue/green 5-cycle and all the other edges (also forming a 5-cycle) are red. In this blue/green 5-cycle there is at least one vertex with two edges of the same color, say blue. This forms a red/blue triangle where the remaining edge is blue or green which is a contradiction to (ii). □

Lemma 2.48. *Every edge-3-coloring of K_8 with a monochromatic triangle is $(2, 2, 2)$-good.*

Proof. Let c be an edge-3-coloring of K_8 with colors red, blue, and green such that there is a red triangle. Then we will show that c is $(2, 2, 2)$-good. Let A be the vertices of the red triangle and $B = V(K_8) \setminus A$ the remaining 5 vertices.

If there exist only red edges in $G[B]$ then we can color A green and B blue. Therefore we can assume that there is a blue or green edge $\{b_1, b_2\}$ and by Lemma 2.47 we can moreover assume that the remaining 3 vertices C do not form a polychromatic triangle.

If C uses only the colors red and green, then we can color A green, C blue, and b_1, b_2 red. If C uses only the colors red and blue, then we can color A blue, C green, and b_1, b_2 red. If C forms a blue/green triangle, then without loss of generality we can assume that $\{b_1, b_2\}$ is green and color A green, C red, and b_1, b_2 blue. □

Proposition 2.49. $s^*(3,2,2) \geq 12$.

Proof. Let c be an edge-3-coloring of K_{11} and our goal is to show that it is $(3,2,2)$-good. Because the Ramsey number $r(4,3)$ is 9 there has to be a red K_4 or a triangle with edges using only the colors blue and green.

Case 1. There is a red K_4.
Let A be the vertices of the red K_4. The remaining 7 vertices have to contain at least two vertex disjoint blue edges (otherwise color A green, the only blue edge red, and the rest blue,). Moreover by Lemma 2.47 we can assume that the remaining 3 vertices do not form a polychromatic triangle. Let us denote the blue edges by e_1, e_2 and the vertex set of the non-polychromatic triangle T.

If T has only blue and green edges then color A blue, e_1 green, e_2 and T red. If T has only red and blue edges then color A blue, e_1, e_2 red, T green. If T has only red and green edges then color A green, e_1, e_2 red, T blue.

Case 2. There is a blue/green triangle.
Let B be the vertices of the triangle and there are 8 remaining vertices in K_{11}. If they form a $(2,2,2)$-critical coloring then it contains no monochromatic triangles by Lemma 2.48 and we can therefore color these vertices with red. For the vertices B we can use blue and green because $s^*(2,2) > 3$. On the other hand if the remaining 8 vertices do not form a $(2,2,2)$-critical coloring then there exists a vertex-coloring using the colors red, blue, and green without monochromatic edges. Furthermore, we can color all vertices in B by red. This coloring yields no red K_3 because for a red K_3 we can only use one of the vertices in B and therefore would need a red edge in the other part. □

Proposition 2.50. $s^*(4,2,2) \geq 15$.

Proof. Let c be an edge-3-coloring of K_{14}, then our goal is to show that c is $(4,2,2)$-good. Because $r(5,3) = 14$, there is either a red K_5 or a

2.4. Cliques—More Colors

blue/green triangle.

First, we assume that there is a blue/green triangle. By Proposition 2.49 there exists a vertex-coloring of the remaining 11 vertices that is $(3, 2, 2)$-good. The extension of this coloring where all vertices from the blue/green triangle are colored red does not create a red K_4.

Second, we assume that there is a red K_5 with the vertex set P. There are at least 3 vertex disjoint blue edges e_1, e_2, e_3 in the remainder (otherwise color P green, the at most 2 vertex disjoint blue edges red, and the rest blue). By Lemma 2.47 we can assume that the remaining triangle T is not polychromatic.

If T uses only the colors red and blue, then color T green, P blue, e_1, e_2, e_3 red. If T uses only the colors red and green, then color T blue, P green, e_1, e_2, e_3 red. If T uses only the colors blue and green, then color T, e_1, e_2 red, P blue, e_3 green. □

Let us summarize the results from this section:

$$\begin{aligned} s^*(2,2,2) &= s(3,3,3) = 8; \\ s^*(2,2,3) &= s(3,3,4) = 12; \\ 15 \leq s^*(2,2,4) &\leq s(3,3,5) \leq 16. \end{aligned}$$

The upper bounds follow from Proposition 2.43 and the lower bounds are proven in the Propositons 2.46, 2.49, 2.50.

> The crocodile is longer than it is green. For a proof, let's look at the crocodile: It is long on the top and on the bottom, but it is green only on the top. Therefore, the crocodile is longer than it is green.
>
> heard from Anita Keszler

Chapter 3

Polychromatic Colorings

We give here first some general framework for polychromatic colorings. Let B be a base set and $\mathcal{F}$ a family of subsets of B. An r-coloring of B can be viewed as a function $\varphi : B \to \{1, 2, \ldots, r\}$. We say that $F \in \mathcal{F}$ is *polychromatic* under φ if it receives all r colors and the coloring φ is called *polychromatic* if every $F \in \mathcal{F}$ is polychromatic under φ. We are interested in the maximum number r of colors such that there exists a polychromatic r-coloring of $\mathcal{F}$. Clearly, this number is upper bounded by the size of any $F \in \mathcal{F}$, and therefore the maximum really exists unless $\mathcal{F} = \emptyset$. Here is a list of some special cases of this problem:

(i) For a hypergraph $H = (V, E)$ and $r = 2$, set $B = V$ and $\mathcal{F} = E$. A polychromatic coloring is here a 2-coloring of the vertices such that each hyperedge receives both colors, i.e., there is no monochromatic hyperedge. Thus, a hypergraph H is polychromatically 2-colorable if and only if it has *property B*.

(ii) For some multigraph G, set $B = E(G)$ and
$$\mathcal{F} = \{\{e : v \in e\} : v \in V(G)\}.$$

A polychromatic coloring is here an edge-coloring such that every vertex receives all colors among its incident edges. The maximization problem is also called the *cover index* of a graph, see [47], and will be discussed in Section 3.1.

(iii) For some plane graph G, set $B = V(G)$ and
$$\mathcal{F} = \{V(f) : f \in F(G)\},$$

where $F(G)$ is the set of all faces in G. The maximization problem is investigated in the subsequent sections starting in Section 3.2.

(iv) For $1 \leq d \leq n$, set $B = V(Q_n)$ and
$$\mathcal{F} = \{S \subseteq V(Q_n) : Q_n[S] \cong Q_d\},$$

where Q_t is the t-dimensional hypercube graph. This case and the corresponding case for edge-coloring is considered in the papers [6, 71] and we will not further discuss it here.

3.1 Polychromatic Edge-Colorings

Definition 3.1. *An edge-r-coloring of a multigraph G is called polychromatic if for all vertices $v \in V(G)$ all r colors appear on the edges incident to v.*

Let G denote a multigraph with a loop e with endpoint $x \in V(G)$ and let G^1, G^2 be two copies of $G-e$ with $x^1 \in V(G^1), x^2 \in V(G^2)$ indicating the special vertex. Then the multigraph $G' = G^1 + G^2 + \{x^1, x^2\}$ is polychromatically edge-r-colorable if and only if G is polychromatically edge-r-colorable. The construction is indicated in Figure 3.1.

By repeatedly applying this procedure we obtain a multigraph G^* with no loops and G^* is polychromatically edge-r-colorable if and only

3.1. Polychromatic Edge-Colorings

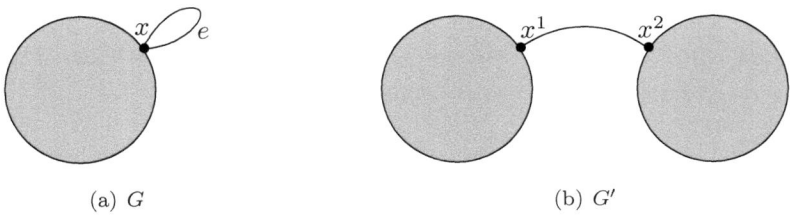

(a) G (b) G'

Figure 3.1: Resolving loops.

if G is polychromatically edge-r-colorable. Thus we can and will assume that all multigraphs in the following have *no loops*.

If there is a polychromatic edge-r-coloring then there is also a polychromatic edge-$(r-1)$-coloring. A polychromatic edge-coloring of G cannot use more than $\delta(G)$ colors where $\delta(G)$ denotes the minimum degree of G.

There is a polychromatic edge-1-coloring for every graph G without isolated vertices. If G has minimum degree 1 then we cannot use more than 1 color. Even cycles are polychromatically edge-2-colorable but graphs containing an isolated odd cycle are not polychromatically edge-2-colorable. The discussion above gives a necessary condition for a multigraph to be polychromatically edge-2-colorable and we show that it is also sufficient:

Proposition 3.2. *A multigraph G is polychromatically edge-2-colorable if and only if the following two conditions hold.*

  (i) *$\delta(G) \geq 2$, and*
 (ii) *G does not contain an odd cycle component.*

Proof. If G contains a vertex of degree at most 1 or and odd cycle component then there is no polychromatic edge-2-coloring which proves that the two conditions are necessary. Sufficiency of the conditions will follow from the following (stronger) statement which will be proven inductively.

Claim 4. *Let G be a multigraph which does not have an odd cycle component. Then there exists an edge-2-coloring of G such that every vertex of degree at least two is polychromatic.*

Proof. The induction proceeds over the number of edges. Without loss of generality G is connected. If there is no cycle at all in G (this covers also the base case for the induction) then $G = T$ is a tree. Let v be a leaf of T and let T_v be the rooted tree corresponding to T with root v. Then we color an edge $\{x, y\}$ with color 1 if $\min\{\text{dist}(v, x), \text{dist}(v, y)\}$ is even and otherwise with color 2. In this way all edges between two levels in T_v have the same color. Every vertex of degree at least two has only one edge to its parent and the other incident edges have the other color. Thus, every vertex of degree at least two is polychromatic.

Let $k > 0$ and let G be a connected multigraph without odd cycle component and with k edges and at least one cycle C. Moreover, we assume that the statement is true for all multigraphs on less than k edges. The cycle C is either (a) an isolated even cycle, (b) an even cycle that is not isolated, or it is (c) an odd cycle that is not isolated.

The alternating edge-2-coloring of C shows that case (a) is trivial. Let $B_1, \ldots, B_s$ be the components of the multigraph obtained by deleting all edges of C in G. We have in case (b) and (c) that $s \geq 1$. Since G is connected, we can choose for each $1 \leq i \leq s$ a vertex $y_i \in V(B_i) \cap V(C)$. If B_j is an odd cycle component, then we can color the edges of B_j alternatingly except that the two edges incident to y_j are colored in color 1. If B_j is not an odd cycle, then inductively there is an edge-coloring of B_j such that every vertex of degree at least two in B_j is polychromatic and, moreover, y_j is incident to an edge in color 1. All vertices of degree at least two in $V(G) \setminus V(C)$ are already polychromatic. Therefore, it is sufficient to check now that there is an edge-coloring of C such that every vertex in C is polychromatic. For case (b), we color the edges in C alternatingly. For case (c), we color the edges in C alternatingly except that the two edges in C incident to y_1 are colored in color 2. Since y_1 has an incident edge in color 1 in B_1, this coloring satisfies the

3.1. Polychromatic Edge-Colorings 63

requirements. □

This finishes also the proof of the proposition. □

As a consequence of the above lemma we have that every graph with minimum degree 3 is polychromatically edge-2-colorable and also every bipartite graph with minimum degree at least two. Our goal is now to prove similar results for more than two colors. To achieve this goal, we show first several simple lemmas with their proof.

Lemma 3.3 ([48, 4, 47]). *Let r be a positive integer. It is possible to color the edges of any bipartite multigraph G by r colors $\{1, \ldots, r\}$, such that for every vertex v of G, the number of edges of each color incident with v are nearly equal. That is for every $i \in \{1, \ldots, r\}$, the number of edges of color i incident with v is either $\lfloor \deg(v)/r \rfloor$ or $\lceil \deg(v)/r \rceil$.*

Proof. First split the vertices of G, if needed, to make its maximum degree at most r: As long as there is a vertex v of G of degree $d > r$, modify it using the following procedure. Define $k = \lceil d/r \rceil$ and replace v by k new vertices $v_1, v_2, \ldots, v_k$, called its descendants. Let $u_1, u_2, \ldots, u_d$ be an arbitrary enumeration of all neighbors of v. For each $i \in [k]$, connect the new vertex v_i with u_j for all j satisfying $(i-1)r < j \leq \min\{d, ir\}$. This process terminates with a bipartite graph in which all degrees are at most r. By König's Theorem (see, for example, [91]) the edges of this graph can be *properly* colored by the r colors. By collapsing all descendants of each vertex v back, keeping the colors of the edges, we obtain an edge-r-coloring of the original graph G satisfying the assertion of the claim. □

Corollary 3.4. *Every bipartite multigraph G has a polychromatic edge-$\delta(G)$-coloring.*

Lemma 3.5. *Every multigraph G contains a spanning bipartite graph $B \subseteq G$ with $\deg_B(v) \geq \lceil \frac{\deg_G(v)}{2} \rceil$ for every $v \in V(G)$.*

Proof. Let B be a maximum edge-cut in G with respect to the number of edges. Assume that there is a vertex $v \in V(G)$ with $\deg_B(v) < \lceil \frac{\deg_G(v)}{2} \rceil$. If we then swap v to the other bipartite set, this would yield another edge-cut with more edges, contradicting the maximality. □

Lemma 3.6. *Every multigraph G has an orientation of its edges such that* $\deg^+(v) \geq \lfloor \frac{\deg(v)}{2} \rfloor$ *for all $v \in V(G)$.*

Proof. We may assume that G is connected. If all degrees in G are even we simply orient it along an Eulerian cycle. Otherwise, define a new graph G' which consists of all vertices of G and a new vertex x and connect all odd degree vertices of G to x. Then all vertices in G' have even degrees and therefore there is an Eulerian cycle in G'. Orient the edges along such an Eulerian cycle and delete the vertex x. Every vertex $v \in V(G)$ with even degree has then exactly $\deg(v)/2$ outgoing edges. Each vertex $v \in V(G)$ with odd degree has either $(\deg(v)+1)/2$ or $(\deg(v)-1)/2$ outgoing edges. □

We are now able to prove a general upper bound for the number of colors in a polychromatic edge-coloring of a multigraph G. This result was also discovered independently by Gupta [47].

Theorem 3.7. *For every multigraph G without isolated vertices there is a polychromatic edge-coloring of G with* $\lfloor \frac{3\delta(G)+1}{4} \rfloor$ *colors.*

Proof. Denote $\delta(G)$ by δ for short. By Lemma 3.5 there is a spanning bipartite subgraph H of G satisfying $\delta(H) \geq \lceil \frac{\delta}{2} \rceil$. Let A_1 and A_2 denote its partite sets. Applying Lemma 3.3 to H with $r = \lfloor \frac{3\delta+1}{4} \rfloor$ results in an edge-coloring χ with the following two properties.

(i) Every vertex v with $\deg_H(v) \geq r$ is polychromatic. Indeed v is incident with at least $\lfloor \deg_H(v)/r \rfloor \geq 1$ edges of each of the r colors.

(ii) For every vertex u with $\deg_H(u) < r$ each color appears at most once on edges incident to u since $\lceil \deg_H(u)/r \rceil = 1$. In other words all edges incident with u have distinct colors.

3.1. Polychromatic Edge-Colorings

Orient the edges of both $G[A_1]$ and $G[A_2]$ according to Lemma 3.6 such that $\deg^+_{G[A_i]}(v) \geq \lfloor \frac{\delta - \deg_H(v)}{2} \rfloor$, for $i = 1, 2$ and all $v \in A_i \subseteq V(G)$. For each vertex $v \in A_i$, color the edges oriented from v to its out-neighbors in $G[A_i]$ with the colors not appearing at the edges of H incident to v (if there are any such colors). Thus, the edges incident with any vertex $v \in V(G)$ are finally colored with

$$\min\left\{\deg_H(v) + \left\lfloor \frac{\delta - \deg_H(v)}{2} \right\rfloor, r\right\} \geq \left\lceil \frac{\delta}{2} \right\rceil + \left\lfloor \frac{\lfloor \frac{\delta}{2} \rfloor}{2} \right\rfloor = \left\lfloor \frac{3\delta + 1}{4} \right\rfloor$$

distinct colors. The inequality follows from the fact that $\deg_H(v) \geq \lceil \frac{\delta}{2} \rceil$. □

Let T_d be the "fat triangle" on the vertices x, y, z with $\lfloor \frac{d}{2} \rfloor$ edges between x and y and $\lceil \frac{d}{2} \rceil$ edges between x and z as well as between y and z (see Figure 3.2).

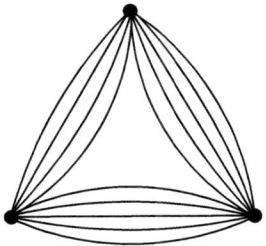

Figure 3.2: Fat triangle T_d for $d = 9$.

It is clear that $\delta(T_d) = d$ and every color class in a polychromatic edge-coloring has to contain at least two edges. This implies that for the number of colors p we can use

$$2p \leq \left\lfloor \frac{d}{2} \right\rfloor + 2\left\lceil \frac{d}{2} \right\rceil,$$

which implies $p \leq \lfloor \frac{3d+1}{4} \rfloor$.

This example shows that the bound of Theorem 3.7 is tight but T_d contains multiedges—necessarily so, since there are better bounds for simple graphs.

Proposition 3.8. *Let G be an r-regular simple graph. Then G is polychromatically edge-$(r-1)$-colorable.*

Proof. Let χ be a proper edge-$(r+1)$-coloring which exists by Vizing's theorem (see, for example, [91]). Every vertex $v \in V(G)$ has one color c_v missing among the colors of its incident edges. Our goal is a new coloring with the first $r-1$ colors which is polychromatic.

Look at the subgraph H of the edges with color r and $r+1$. It is a union of paths and cycles. We can orient them such that each path/cycle is an oriented path/cycle. The vertices which are not polychromatic for the first $r-1$ colors are exactly the vertices of degree 2 in H. For every vertex v with $c_v \in [r-1]$ we take the outgoing edge in H and recolor it with the color c_v. Any completion of this coloring to all the edges yields an polychromatic edge-$(r-1)$-coloring. □

Theorem 3.9 (Gupta [47]). *For any multigraph G and any $k \in \mathbb{N}$, there exists an edge-k-coloring of G such that for every vertex x the number of distinct colors appearing at the edges incident to x is at least*

(i) $\min\{k - m_G(x), \deg(x)\}$,   *if* $\deg(x) \leq k$;
(ii) $\min\{k, \deg(x) - m_G(x)\}$,   *if* $\deg(x) \geq k$.

where $m_G(x)$ is the maximum number of edges between x and any of its neighbors.

By applying Theorem 3.9 with $k = \delta(G)$ to a simple graph G we obtain a generalization of Proposition 3.8.

Corollary 3.10. *A simple graph G is polychromatically edge-$(\delta(G)-1)$-colorable.*

Proposition 3.2 gives a complete, polynomially testable characterization for graphs which are polychromatically edge-2-colorable. As not uncommon in complexity theory, the case $r = 3$ is then already NP-complete as well as for any other fixed $r \geq 3$.

Theorem 3.11 (Hoyler [51], Leizhen and Ellis [23]). *For any $r \geq 3$, it is* NP-*hard to decide whether a simple r-regular graph is properly edge-r-colorable.*

For r-regular graphs the notion of proper edge-r-coloring and polychromatic edge-r-coloring is equivalent. Thus the above theorem tells us that also polychromatic edge-3-colorability is NP-hard.

Corollary 3.12. *For any $r \geq 3$, it is* NP-*hard to decide whether a simple r-regular graph is polychromatic r-edge colorable.*

3.2 Colorings of Plane Multigraphs

We will work mostly with plane multigraphs for the remaining of this chapter, i.e., we assume that the graph is embedded in the plane without crossings. Denote the set of faces of a plane multigraph G by $F(G)$.

Definition 3.13. *For a vertex k-coloring of G we say that a face $f \in F(G)$ is* polychromatic *if all k colors appear on the vertices of f. A vertex k-coloring of G is called* polychromatic *if every face (also the outerface) of G is polychromatic. The* polychromatic number *of G, denoted by $p(G)$, is the largest number k of colors such that there is a polychromatic vertex k-coloring of G.*

Note that a polychromatic coloring does not have to be proper, i.e., it is possible that both endpoints of an edge receive the same color.

Moreover, we want to remark that we can neglect loops: If G is a plane multigraph with a loop around x, then we can divide the graph into the interior G_{in} and the exterior G_{ext} of the loop both including the vertex x but not the loop itself.

It is easy to combine a vertex-coloring of G_{in} and G_{ext} and therefore we get $p(G) = \min\{p(G_{in}), p(G_{ext})\}$. Thus, we will assume in the following that G is a plane multigraph *without loops*.

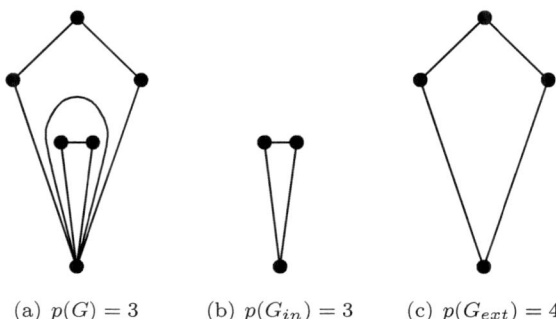

(a) $p(G) = 3$ (b) $p(G_{in}) = 3$ (c) $p(G_{ext}) = 4$

Figure 3.3: Example. $p(G) = \min\{p(G_{in}), p(G_{ext})\}$

The *size* of a face $f \in F(G)$ is the number of vertices on its boundary. For a plane graph G, let $g(G)$ denote the size of the smallest face in G. A face of size s in a plane graph is sometimes also called an *s-face*. Define

$$p(g) = \min\{p(G) \mid G \text{ plane graph}, g(G) = g\}.$$

It is clear that $p(g(G)) \leq p(G) \leq g(G)$ for every plane graph G. By adding vertices inside a face of G, we can increase the size of the smallest face without decreasing the maximum number of colors one can use in a polychromatic coloring. Thus, the function $p(g)$ is non-decreasing, i.e., for $g \leq g'$ we have $p(g) \leq p(g')$.

If $g(G) = 1$, then G contains only one vertex and therefore $p(1) \leq 1$. If $g(G) = 2$, then G contains either multiple edges or only two vertices. The graph G' depicted in Figure 3.4 shows that also $p(2) \leq 1$.

Figure 3.4: Graph G' with $g(G') = 2$ and $p(G') = 1$.

3.2. Colorings of Plane Multigraphs

It is well-known that every plane *simple* graph can be polychromatically 2-colorable, see for example [13], [66], [12]. We will generalize this result to plane multigraphs G with $g(G) \geq 3$.

A *triangulation* of a plane multigraph G is obtained from G by adding edges such that every face is a 3-cycle.

Lemma 3.14. *Let G be a plane multigraph with $g(G) \geq 3$. There exists a triangulation H of G.*

Proof. Without loss of generality we can assume that G is connected. If there are two vertices in a face $f \in F(G)$ that are not connected inside f then we can add an edge between them inside f and receive again a plane multigraph with two new faces f_1, f_2. The size of the new faces f_1, f_2 is smaller than the size of the old face f. Therefore, the process stops at some point where all faces are cliques. If we start with a plane multigraph G with $g(G) \geq 3$, then we will not have 1- or 2-faces after the process. Every face is an outerplanar graph and therefore we cannot have a K_4. The only possibility remaining is that all faces are K_3, i.e., 3-cycles, which is then the desired triangulation. $\square$

Theorem 3.15. *Every plane multigraph G with $g(G) \geq 3$ is polychromatically 2-colorable.*

Proof. Triangulate the graph G by adding edges, resulting in a new graph H where each face (also the outerface) is a 3-cycle. The dual graph H^* of H is then 3-regular. Moreover, H^* is 2-edge connected: every minimal edge-cut in H^* correspond to a cycle in H, and since H has no loop, there is no cut-edge in H^*. By Petersen's Theorem (see, for example, [91]), there exists a perfect matching M in H^*. After deleting the edges of H corresponding to those of M, the remaining graph H' has only faces of size 4. Therefore there is no odd cycle in H' and hence H' is bipartite. Thus, there is a proper vertex 2-coloring of H', which is a polychromatic vertex 2-coloring of H and hence also of G. $\square$

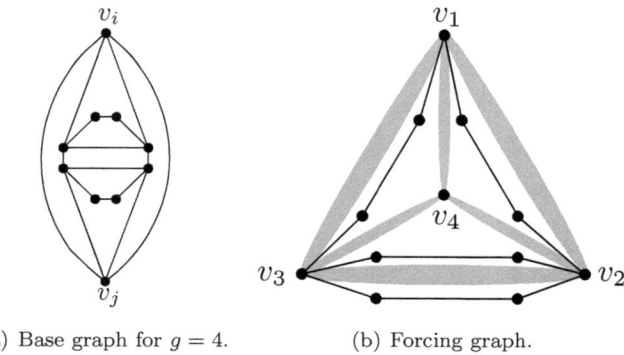

(a) Base graph for $g = 4$.   (b) Forcing graph.

Figure 3.5: Graph G with $p(G) = 2$ and $g(G) = 4$.

A planar embedding of K_4 has $g(K_4) = 3$ and $p(K_4) = 2$. For $g = 4$ consider Figures 3.5(a) and (b) which illustrate the construction of a graph G. The graph G equals the *forcing graph* (see Figure 3.5(b)) where each of the six shaded edges $\{v_i, v_j\}$ is replaced by a copy of the base graph (see Figure 3.5(a)). Clearly, $g(G) = 4$. It is easy to check that the following holds.

In any polychromatic 3-coloring of a base graph (see Figure 3.5(a)) the vertices v_i and v_j are colored with distinct colors.

Thus from the fact that K_4, the graph underlying the forcing graph, is not properly 3-colorable, it follows that $p(G) \leq 2$.

Therefore, we have

$$p(1) = 1, p(2) = 1, p(3) = 2, p(4) = 2,$$

and $p(5) \geq 2$ but we think that the true value should be 3.

Conjecture 3.16. *Every plane graph G with $g(G) \geq 5$ is polychromatically 3-colorable.*

We proceed in the next subsections by giving lower and upper bounds for plane graphs G with $g(G) \geq 5$, more precisely, we will show the

3.2. Colorings of Plane Multigraphs

following bounds for $g \geq 5$

$$\left\lfloor \frac{3g-5}{4} \right\rfloor \leq p(g) \leq \left\lfloor \frac{3g+1}{4} \right\rfloor. \tag{3.1}$$

Note that the set $\{\lfloor \frac{3g-5}{4} \rfloor, \ldots, \lfloor \frac{3g+1}{4} \rfloor\}$ contains at most 3 integers.

3.2.1 The Lower Bound

We first present several small lemmas which will be needed for the proof of the lower bound. An incidence is a pair (v, f) where v is vertex and f a face such that v is on the boundary f.

Lemma 3.17. *Let G be a plane graph, let $\emptyset \neq F' \subseteq F(G), \emptyset \neq V' \subseteq V(G)$ and let $i(V', F')$ denote the number of incidences between F' and V'. Then $i(V', F') \leq 2|F'| + 2|V'| - 3$.*

Proof. Define the incidence graph H of $V' \subseteq V(G)$ and $F' \subseteq F(G)$ by $V(H) = F' \cup V'$ and $\{f, v\} \in E(H)$ for $v \in V', f \in F'$ if and only if v is on the boundary of f in G. It is easy to see that H is planar, simple and bipartite. From Euler's Formula and the fact that H is simple and triangle-free it follows that H contains at most $2V(H) - 4$ edges, provided that H contains at least three vertices. In this case we conclude that $i(V', F') = |E(H)| \leq 2(|V'| + |F'|) - 4$. On the other hand if $|V(H)| = 2$ and H contains one edge, then $i(V', F') = 2(|V'| + |F'|) - 3 = 1$. □

The following result is well known (see, for example, [64], Theorem 2.4.2). For completeness, we include a proof sketch.

Lemma 3.18. *Let $A \in \{0,1\}^{m \times n}$ be a matrix with entries $a_{i,j}$ for $i \in [m], j \in [n]$. The following two statements are equivalent:*

(i) *There is a matrix $C \in \{0,1\}^{m \times n}$, $C \leq A$ (that is $c_{i,j} \leq a_{i,j}$ for all $i \in \{1, \ldots, m\}$ and all $j \in \{1, \ldots, n\}$) such that every row in C contains at least q 1's and every column in C contains at most r 1's.*

(ii) *For every $M \subseteq \{1,\ldots,m\}$ and every $N \subseteq \{1,\ldots,n\}$, $\sum_{i\in M, j\in\{1,\ldots,n\}\setminus N} a_{i,j} \geq q|M| - r|N|$.*

Proof. Define a network with vertices $s, t, r_1, \ldots, r_m, c_1, \ldots, c_n$ as follows. Connect the source s with all vertices r_i with edges having capacity q, connect r_i with c_j with edges having capacity $a_{i,j}$, and connect all c_j to the sink t with edges having capacity r. If condition (i) holds, then we can also assume that there exists such a matrix C where in every row there are exactly q 1's. Thus there exists a flow of value mq if and only if (i) holds. It is easy to show that all cuts have size at least qm if and only if condition (ii) holds. This implies the statement by using the MaxFlow-MinCut Theorem. □

Corollary 3.19. *Let G be a plane graph with $g(G) = g$. For each face $f \in F(G)$ we can assign $g - 2$ vertices that lie on its boundary such that no vertex is assigned to more than two faces.*

Proof. Let $A = (a_{f,v})_{f\in F, v\in V} \in \{0,1\}^{|F|\times|V|}$ be the face-vertex incidence matrix of G where $F = F(G)$ and $V = V(G)$. That is $a_{f,v} = 1$ if and only if vertex v is contained in face f. We want to show that there is a matrix $C \in \{0,1\}^{|F)|\times|V|}$ such that $C \leq A$, in every row of C there are at least $(g-2)$ 1's, and in every column of C there are at most two 1's.

By Lemma 3.18 with $q = g - 2$ and $r = 2$ it is sufficient to show that for every $F' \subseteq F, V' \subseteq V$, $\sum_{f\in F', v\in V\setminus V'} a_{f,v} \geq (g-2)|F'| - 2|V'|$.

Henceforth we obtain

$$\sum_{f\in F', v\in V\setminus V'} a_{f,v} = \sum_{f\in F', v\in V} a_{f,v} - \sum_{f\in F', v\in V'} a_{f,v}$$
$$\geq g|F'| - \sum_{f\in F', v\in V'} a_{f,v}$$
$$\geq g|F'| - 2|F'| - 2|V'|,$$

where the last inequality follows from Lemma 3.17 in case both V' and F' are nonempty, and is trivial if at least one of them is empty. □

3.2. Colorings of Plane Multigraphs

Theorem 3.20. *For $g \geq 5$*

$$p(g) \geq \left\lfloor \frac{3g-5}{4} \right\rfloor.$$

Proof. Let $G = (V, E)$ be a plane graph with $g(G) = g$. By Corollary 3.19 we can assign $g-2$ vertices from its boundary to every face of G such that no vertex is assigned to more than two faces of G. Define an auxiliary multigraph H, with $V(H) = F(G) \cup \{x, y\}$, where x, y are two additional vertices. For every vertex $v \in V(G)$ define an edge of H, which we call the v-edge, as follows. If v is assigned to two distinct faces f_1 and f_2 then the v-edge is $\{f_1, f_2\}$. If it is assigned only to one face f, the v-edge is $\{f, x\}$, and if it is not assigned to any face, then the v-edge is $\{f, y\}$. In addition, add $g-2$ (multi)edges to H connecting x and y to ensure that all degrees in H are at least $g-2$. Thus, H is a loopless multigraph with minimum degree at least $g-2$. By Theorem 3.7 with $d = g-2$ we can color the edges of H with $p = \lfloor \frac{3(g-2)+1}{4} \rfloor = \lfloor \frac{3g-5}{4} \rfloor$ colors such that every vertex $f \in V(H)$ is incident with edges of all p colors.

Define a vertex-coloring of G by coloring every vertex $v \in V(G)$ by the same color as that of the v-edge. This clearly gives a coloring in which every face $f \in F(G)$ is polychromatic, as needed. $\square$

The above proof is constructive, i.e., one can find in polynomial time a polychromatic coloring of G with $\lfloor \frac{3g-5}{4} \rfloor$ colors.

3.2.2 The Upper Bound

Theorem 3.21. *For $g \geq 5$,*

$$p(g) \leq \left\lfloor \frac{3g+1}{4} \right\rfloor.$$

Proof. Define the graph G_g as depicted in Figure 3.6. For g even set $k = l = \frac{g}{2}$ and for g odd set $k = \frac{g+1}{2}$ and $l = \frac{g-1}{2}$. Inside the small triangle and outside the big triangle add a path of $g-2$ new vertices as indicated by the dashed arcs. Then $g(G_g) = g$. Note that the vertices of the three faces of G_g that contain no dashed arcs are $W := \{u_1, u_2, \ldots, u_k, w_1, w_2, \ldots, w_k, v_1, v_2 \ldots, v_l\}$, and none of these vertices lies in all three faces. This implies:

In every polychromatic coloring of G_g, every color appears on at least two vertices in the set W.

Therefore

$$2p(G_g) \leq |W| = 2k + l = \begin{cases} 3k, & \text{if } g \text{ is even.} \\ 3k - 1, & \text{if } g \text{ is odd.} \end{cases}$$

$$= \begin{cases} \frac{3g}{2}, & \text{if } g \text{ is even.} \\ \frac{3g+1}{2}, & \text{if } g \text{ is odd.} \end{cases}$$

In both cases we have $p(G_g) \leq \lfloor \frac{3g+1}{4} \rfloor$. Furthermore, this is a construction of a *simple* plane graph. $\square$

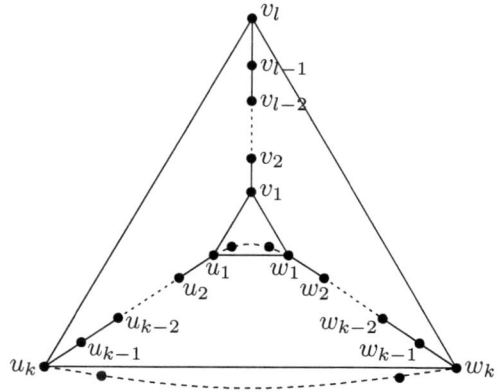

Figure 3.6: Graph G_g with $g(G_g) = g$ and $p(G_g) \leq \lfloor \frac{3g+1}{4} \rfloor$.

3.3 Special Cases of Plane Graphs

There are special cases of plane graphs with better bounds for the polychromatic number.

3.3.1 Triangulations

For every triangulation G it holds that $2 \leq p(G) \leq 3$. The following simple characterization of triangulations G with $p(G) = 3$ is a consequence of an old result of Heawood [49].

Theorem 3.22. *Let G be a triangulation. The following two statements are equivalent:*

(i) $p(G) = 3$, *and*
(ii) G *is Eulerian, i.e., the degree of every vertex in G is even.*

The following two results immediatly imply Theorem 3.22.

Theorem 3.23 (Kempe [57], Heawood [49, 86]). *The vertices of a triangulation G are properly 3-colorable if and only if G is Eulerian.*

Lemma 3.24. *Let G be a triangulation. Then the following are equivalent:*

(i) G *is polychromatically 3-colorable.*
(ii) G *is properly 3-colorable.*

Proof. A triangle is properly 3-colored if and only if its three vertices have the three different colors. Also, a triangle is polychromatically 3-colored if and only if its three vertices have the three different colors. Thus these two notions are equivalent for triangulations. □

3.3.2 Graphs with Only Even Faces

We will show that multigraphs with only faces of even sizes are polychromatically 3-colorable by using the same statement for simple plane graphs.

Theorem 3.25 (Hoffmann and Kriegel [50]). *Let G be a graph which is 2-connected, bipartite, and simple, and plane. Then we can add edges to G to obtain a triangulation such that the degree of every vertex is even. Moreover, this triangulation can be found in polynomial time.*

Theorem 3.26. *Let G be a 2-connected, plane multigraph with even faces only and $g(G) \geq 4$, then there exists a polychromatic 3-coloring of G that is proper as well, i.e., no edge is monochromatic. Moreover, such a coloring can be found in polynomial time.*

Proof. We prove the statement by induction on the number of multiple edges of G. First, we assume that G is simple. Every cycle in G has even length (i.e., G is bipartite) because G is required to be 2-connected and has only even faces. The statement follows after applying Theorem 3.25 and Theorem 3.22.

Next, we assume that G has some multiple edges. Let $x, y \in V(G)$ and $e_1, e_2 \in E(G)$ where both e_1 and e_2 connect x and y. The edges e_1, e_2 build a cycle of length two and therefore they divide the plane into two parts. Let V_1 be the vertices inside e_1, e_2 (including x, y) and V_2 the vertices outside e_1, e_2 (including x, y). Since $g(G) \geq 4$, we can conclude that $V_i \supsetneq \{x, y\}$, for $i = 1, 2$. Define $G_1 = (V_1, E(G[V_1]) \setminus \{e_2\})$ and $G_2 = (V_2, E(G[V_2]) \setminus \{e_1\})$. These two graphs are plane, 2-connected with even faces only, $g(G_1), g(G_2) \geq 4$, and each G_i contains less multiple edges than G. There exists inductively a polychromatic 3-coloring of G_i, $i = 1, 2$, such that no edge is monochromatic. In particular the coloring of G_1 and the coloring of G_2 assigns distinct colors to x and y. Thus we can permute the colors of one coloring such that the colors of x and y agree in the colorings of G_1 and of G_2. This yields a 3-coloring of G which fulfills the condition in the statement. □

3.3.3 Outerplanar Graphs

Another simple case is when the multigraph G is outerplanar (i.e., all vertices lie on the outerface). The size of the smallest face is then equal

3.3. Special Cases of Plane Graphs

to the length of the smallest cycle (girth of G) unless G is a forest. We show that the trivial upper bound $p(G) \leq g(G)$ is tight for outerplanar graphs G with $g(G) \geq 3$.

Theorem 3.27. *Let G be an outerplanar graph with $g = g(G) \geq 3$. Then there exists a polychromatic coloring of G with g colors that is also proper, i.e., no edge is monochromatic.*

Proof. We prove this result by induction on the number of faces. If we have only one face, then the graph G is a forest and clearly we can polychromatically color every forest with $|V(G)| = g(G)$ many colors such that no edge is monochromatic. Let us assume now that G has more than one face. Obviously, it is sufficient to find a g-coloring of the vertices of G such that all bounded faces are polychromatic and no edge is monochromatic. The outerface will by the outerplanarity automatically be polychromatic since all vertices lie on the outerface. Also we can assume without loss of generality that G is connected and has no cut-vertex. Otherwise color the 2-connected components separately and combine the coloring (maybe rename the colors in each component correspondingly).

It is well-known that the dual graph G^* without the outerface forms a forest; and since G is 2-connected, G^* is connected, and so G^* forms a tree. Every tree has at least two leaves. Choose f_0 as a face corresponding to a leaf in the tree with maximal size. Let G' be the graph obtained from G after deleting all vertices incident to only f_0 and the outerface. Then G' is an outerplanar graph and has one fewer face than G. Moreover, since f_0 was choosen to be of maximal size, we have $g(G') = g(G)$. By the induction hypothesis we can color G' polychromatically with g colors such that no edge is monochromatic.

Finally, add f_0 again to G'. There is exactly one edge $e_0 \in E(G')$ which is on the boundary of the face f_0, i.e., e_0 is the edge between f_0 and its parent. The intersection of the vertices of f_0 and $V(G')$ are exactly the endpoints z_1, z_2 of e_0. For simplicity, assume that z_1 has color 1 and z_2 has color 2. Let $z_3, \ldots, z_k$ be the other vertices of f_0 such that $z_1, z_2, z_3, \ldots z_k$ is the clockwise or counterclockwise order in that face. Extend the coloring of f_0 to $1, 2, \ldots, g, g-1, g, g-1, \ldots$. The face f_0 will then be polychromatic (because $k \geq g$) and no edge of f_0 will be monochromatic (because $g \geq 3$). □

The graph G_2 from Figure 3.4(b) shows an outerplanar graph with $g(G_2) = 2$ which is not polychromatically 2-colorable.

3.4 Connection to Guarding Problems

Polychromatic colorings are related to a combinatorial version of *guarding problems* on graphs. In general, guarding problems ask for a small set of points (guards) that *see* a given input domain, for example a polygon, a terrain, or a plane graph. If we consider guarding a plane graph G, then G is guarded if every face of G is guarded. If all faces are convex, then every vertex on the boundary of a face sees the complete face. If the faces are not convex, more guards might be necessary. Certainly a guard cannot see the entire unbounded face, hence the outerface is usually not required to be guarded. A combinatorial variant of this problem is the following: Find the smallest set of vertices S of G such that every face is incident to (at least) one of the vertices in S. Clearly each color class in a polychromatic coloring is a guarding set, that is, the vertices in each color class jointly *guard* the graph G. From now on we use "guard" in this combinatorial sense and also require the unbounded face to be guarded.

In [13] it is shown that one can guard any plane graph on n vertices with no faces of size 1 or 2 by $\lfloor \frac{n}{2} \rfloor$ guards. This clearly follows from the fact that $p(G) \geq 2$ for any such graph. Similarly, a simple consequence of Theorem 3.20 is the following:

Corollary 3.28. *Every plane graph G with $g(G) = g$ can be guarded with at most $\frac{n}{\lfloor (3g-5)/4 \rfloor} \leq \frac{4n}{3g-8}$ guards.*

Proof. By Theorem 3.20, G admits a polychromatic $\lfloor \frac{3g-5}{4} \rfloor$-coloring. Place guards on the vertices of the smallest color class which is of size at most $\frac{n}{\lfloor \frac{3g-5}{4} \rfloor} \leq \frac{4n}{3g-8}$. Because the coloring is polychromatic each face is incident to at least one guard and the statement follows. □

3.5 Complexity Results for Plane Graphs

Theorem 3.22 immediately implies a polynomial time algorithm to decide whether a triangulation admits a polychromatic 3-coloring.

For general plane graphs G we show that the decision problem whether G is polychromatically 3-colorable is hard and also for polychromatic 4-colorings.

Theorem 3.29. *The decision problem whether a plane graph is polychromatically k-colorable is*

(i) *in P, if $k = 2$, and*
(ii) *NP-complete, for $k = 3, 4$.*

Moreover, we consider the decision problem whether a 2-connected plane graph with faces of size restricted to a set of integers admits a polychromatic 3-coloring. We achieve an almost complete characterization of such sets of integers (face sizes) for which the corresponding decision problem is NP-complete and for the others it is in P.

It can be checked in polynomial time whether a k-coloring is polychromatic, and therefore the problem is in NP. Every plane graph is polychromatically 1-colorable. Thus the decision problem for $k = 1$ is trivial in the sense that the answer for every instance is always "Yes".

Next, we turn our focus to polychromatic 2-colorings and prove Theorem 3.29(i). At this point, it is worth to remind ourselves that every plane graph G with $p(G) < 2$ contains a face of size at most two.

Proposition 3.30. *There is a polynomial time algorithm to decide whether a given plane graph is polychromatically 2-colorable.*

Proof. We call a CNF-formula F *-planar if its literal-clause incidence graph H is planar. Note that this differs from the common notion of a planar CNF-formula, where one assumes that the literal-clause incidence graph H together with a cycle connecting the positive literals and together with edges between the corresponding positive and negative literals is required to be planar.

A vertex-coloring of a plane graph is 2-polychromatic if no face is monochromatic. We can associate with one color the logic predicate 'true' and with the other color 'false' and interpret the vertices as boolean variables. Then we add a clause-vertex to each face and connect it to its incident variable-vertices. By this we get a *-planar CNF-formula (where all variables occur only as positive ones).

Deciding whether a plane graph is polychromatically 2-colorable is equivalent to deciding whether the corresponding planar* CNF-formula is not-all-equal satisfiable (*-PLANAR-NAE-SAT).

In [67] it is shown that PLANAR-NAE-3-SAT is in P by a reduction to PLANAR-MAX-CUT. The reduction in fact holds also for PLANAR*-NAE-SAT. A well known reduction works to shorten the clauses of a planar (and planar*) formula to length 3, whilst preserving not-all-equal satisfiability and planarity. We briefly sketch this reduction which is illustrated in Figure 3.7. A clause c of length $k > 3$ is replaced by two clauses c_1, c_2 of length 3 and $k-1$, respectively. A new

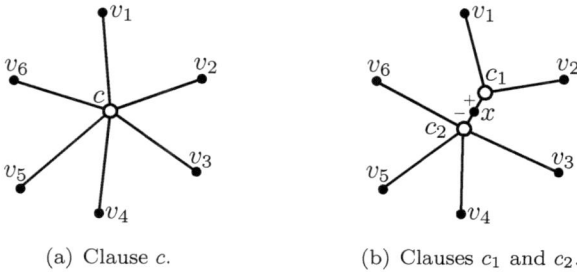

(a) Clause c.   (b) Clauses c_1 and c_2.

Figure 3.7: Reducing PLANAR-NAE-SAT to PLANAR-NAE-3-SAT.

3.5. Complexity Results for Plane Graphs

variable x occurs positive in c_1 and negative in c_2. Placing the new variable and clauses as in Figure 3.7 preserves planarity and not-all-equal satisfiability. □

In the following we want to show hardness results for polychromatic 3- and 4-colorings, by constructing reductions from proper 3-colorability of plane graphs. We start by proving Theorem 3.29(ii) for $k = 3$.

Proposition 3.31. *It is NP-hard to decide whether a given plane simple graph is polychromatically 3-colorable.*

Proof. It has been shown in [87] that deciding whether a plane simple graph is properly 3-colorable is NP-hard. Given a plane simple graph G we construct in polynomial time a plane simple graph G' such that G is properly 3-colorable if and only if G' is polychromatically 3-colorable.

For every edge $e = \{u, v\} \in E(G)$ we add a new vertex y_e inside one of the two faces and connect it with u and v. Thus every edge $e \in E(G)$ is now contained in a triangle. Furthermore, for every face $f \in F(G)$ select a vertex x incident to f. Then add a new vertex x_f into the interior of f and connect x and x_f by an edge. The resulting graph G' is simple. In every polychromatic 3-coloring of G' the edges $E(G)$ are not monochromatic, and every proper 3-coloring of G can be extended to a polychromatic 3-coloring of G' by using the extra vertices x_f. Thus G' is polychromatically 3-colorable if and only if G is properly 3-colorable. □

We will refine Proposition 3.31 by restricting on plane graphs with only faces of given sizes. To do so we will restrict on 2-connected graphs. One reason is that a graph G is properly k-colorable if and only if all its 2-connected components are properly k-colorable and the maximal 2-connected components (block-cutvertex graph) of a graph can be computed in polynomial time by using a depth-first-search. Thus it follows that proper k-colorability is also NP-hard restricted on 2-connected graphs for $k \geq 3$. Another reason is that any face in a 2-connected plane graph is a cycle and therefore there are no artifacts such as dangling paths.

Let L denote some set of positive integers. We define the following two decision problems.

L-PLANE-PROPER-k-COLORABILITY:
Given: A plane 2-connected graph G where the size of each face of G is in L.
Question: Does there exist a proper k-coloring of $V(G)$?

L-PLANE-POLY-k-COLORABILITY:
Given: A plane 2-connected graph G where the size of each face of G is in L.
Question: Does there exist a polychromatic k-coloring of $V(G)$?

In case we do not impose any restriction on the sizes of the faces in G we omit the set L.

Let f be a face of a plane graph G and $L \subseteq \mathbb{N}$. We say a plane graph G' is an L-*extension* of f if G' is a plane graph containing G and some new vertices $V' \neq \emptyset$ and some new edges $E' \neq \emptyset$ (thus also some new faces) such that

(i) the new vertices V' and the new edges E' are contained in the interior of f,

(ii) every new edge of E' is incident to at most one old vertex $v \in V(f)$, and

(iii) the size of any new face is contained in L.

An extension is called 2-*degenerate* if there is an order $v_1, \ldots, v_k$ of the new vertices V', such that the $d_{G'[V(G) \cup \{v_1,\ldots,v_i\}]}(v_i) \leq 2$, for all $i \in \{1, \ldots, k\}$. It is easy to observe now that the following is true.

> Let G' be a 2-degenerate extension of f of G. Any proper 3-coloring of G can be extended to a proper 3-coloring of G',

3.5. Complexity Results for Plane Graphs

i.e., it preserves proper-3-colorability.

Lemma 3.32. *Let $k \geq 3$. Every k-face f of a plane 2-connected graph G has a $\{3,4,5\}$-extension G' in G that is 2-degenerate and 2-connected.*

Proof. The statement is trivial for $k = 3, 4$, or 5. Therefore assume $k \geq 6$ and assume that the statement is true for every smaller k. Let $x_1, \ldots, x_k$ be the vertices of f in clockwise order. Let H be the graph obtained from G by adding a vertex y in the interior of f and connecting y with x_1 and x_4. Then H has a 5-face and a $(k-1)$-face. By induction assumption we can extend the $(k-1)$-face to 3-,4-,5-faces such that the extension is 2-degenerate and 2-connected. Together this yields a $\{3,4,5\}$-extension G' that is 2-degenerate and 2-connected. □

Lemma 3.33. *Every 5-face f of a plane 2-connected graph G has a $\{3,4\}$-extension G' that is 2-connected and moreover G is properly 3-colorable if and only if G' is properly 3-colorable.*

Proof. First note that every 5-face forms a 5-cycle due to the assumption that G is 2-connected. We extend each 5-face f by the construction depicted in Figure 3.8. Specifically, let f be a 5-face and let $v_1, v_2, \ldots, v_5$ be the five vertices of f. We add two copies of P_2 (the path of length two) with vertices u, v, w, $P' : u', v', w'$ and $P'' : u'', v'', w''$ by identifying both u' and u'' with v_1, w' with v_3 and w'' with v_4. Further we connect v' with v''. This yields the $\{3,4\}$-extension G' of G which also is 2-connected.

Trivially every proper 3-coloring of G' is also a proper 3-coloring of G, because G is a subgraph of G'.

On the other hand, every proper 3-coloring of G can be extended to a proper 3-coloring of G' (i.e., it preserves proper-3-colorability). Any proper 3-coloring χ of the 5-face has an extension to a proper 3-coloring of G': We can assume that $\chi(v_1) \neq \chi(v_4)$. Color v' with $\chi(v_4)$ and color v'' with the third color not appearing on any of the neighbors of v''. □

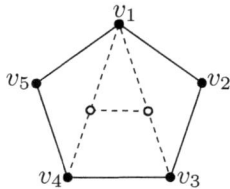

Figure 3.8: Fill graph for 5-faces.

Lemma 3.34. *Let G be a plane 2-connected graph.*

(i) *Let $s \geq 4$. Every 4-face of G has a 2-degenerate $\{3, s\}$-extension G' such that G' is 2-connected as well.*

(ii) *Let $t \geq 5$ odd. Every 3-face and every 4-face has a 2-degenerate $\{t\}$-extension G' such that G' is 2-connected.*

Proof. (i) For $s = 11$ the extension is drawn in Figure 3.9(a) and it should be clear how to obtain a similar construction for arbitrary s.

(ii) In Figure 3.9(b) an extension of a 3-face into 4-faces and 9-faces is shown. The 4-faces can be extended into 9-faces as shown in Figure 3.9(c). Together this gives the extensions for the case $t = 9$. Again the general case should be clear. □

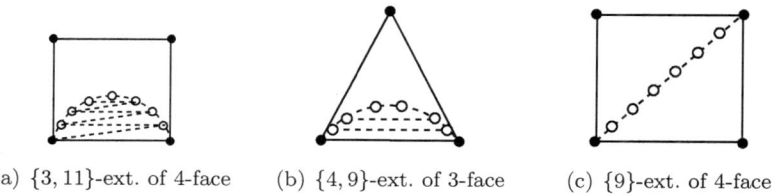

(a) $\{3, 11\}$-ext. of 4-face (b) $\{4, 9\}$-ext. of 3-face (c) $\{9\}$-ext. of 4-face

Figure 3.9: 2-degenerate extensions of faces

This leads to the following complete characterization of the complexity of *L*-PLANE-PROPER-3-COLORABILITY:

3.5. Complexity Results for Plane Graphs

Corollary 3.35. *L-PLANE-PROPER-3-COLORABILITY*

(i) ... *is in* P *for* $L = \{2, 3\}$.
(ii) ... *is trivial provided that L contains only even numbers.*
(iii) ... *is* NP-*complete provided there is $t \in L$ with $t \geq 5$ odd.*
(iv) ... *is* NP-*complete provided $3 \in L$ and there is $s \in L$ with $s \geq 4$.*

Proof. First observe that we can assume that G contains no face of size two, since deleting one edge from a 2-face does neither change the size of any other face of G nor does it yield any cut-vertex.

(i) The only case left is $L = \{3\}$, i.e., triangulations. Theorem 3.22 provides a polynomial time checkable criterion for 3-colorability of triangulations.

(ii) The graphs are bipartite because any cycle has even length. Therefore there is a proper 2-coloring which is also a proper 3-coloring.

(iii), (iv) Using Lemma 3.32, Lemma 3.33, and Lemma 3.34 we can extend every plane 2-connected graph to a graph only having faces of the given size such that the proper 3-colorability and 2-connectedness is preserved. Thus the restricted proper 3-colorability problem on plane, 2-connected graphs is as hard as the non-restricted one. □

Note here that every proper 3-coloring of an odd face is a polychromatic 3-coloring as well. For even faces some special care has to be taken.

Lemma 3.36. *Let $s \geq 4$ even and let C be an s-cycle embedded in the plane. Then there exists an $\{s\}$-extension C' of C such that any proper 3-coloring of C can be extended to a 3-coloring of C' such that every bounded face is polychromatic. Moreover, C' is 2-connected as well.*

Proof. First, we consider the case for $s = 4$. We "fill" C by substituting it with a copy of the graph in Figure 3.10(a). Let v_1, v_2, v_3, v_4 be the four consecutive vertices of C. We identify v_i with the copy of the vertex u_i for $i \in \{1, \ldots, 4\}$. The resulting subgraph is polychromatically

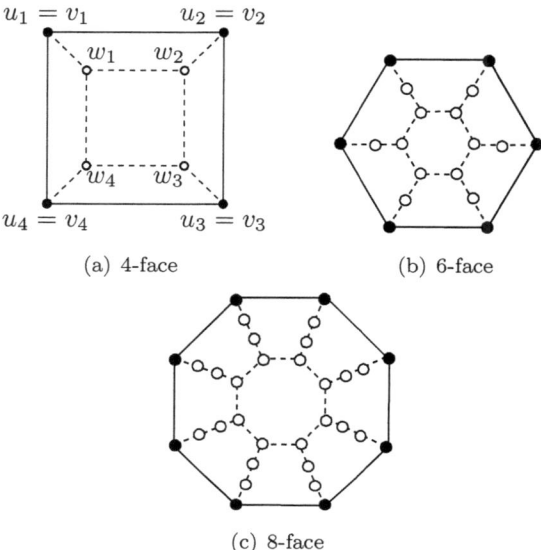

Figure 3.10: Fill graphs.

3-colorable if f is properly 3-colorable. To see this, we fix a proper 3-coloring χ of f. Suppose first that all three colors appear on the vertices of f. Without loss of generality we can assume that $\chi(v_1) = 1, \chi(v_2) = 2, \chi(v_3) = 3$ and $\chi(v_4) = 2$. Then, for instance, coloring the copies of w_1 by 3, w_2 by 2, w_3 by 1 and w_4 by 2 extends χ to a 3-coloring of the new vertices in f such that each of the five new faces inside f is polychromatic. Suppose now that only two distinct colors appear on the vertices of C, say $\chi(v_1) = \chi(v_3) = 1$ and $\chi(v_2) = \chi(v_4) = 2$. We can extend χ to a polychromatic 3-coloring including the new vertices in C as follows. Color w_1 by 3, w_2 by 2, w_3 by 3 and w_4 by 1. Again the five new faces inside C are polychromatic.

The case $s \geq 6$ is even simpler and we will only sketch it here. We use a similar construction as for the previous case (see Figure 3.10(b),(c) for the cases $s = 6$ and $s = 8$). The claim is now that every proper 3-coloring can be extend to a polychromatic 3-coloring inside that face.

3.5. Complexity Results for Plane Graphs

The new faces incident to the original boundary have a non-monochromatic edge already colored. For each such face f we can assign one incident vertex x_f that is not incident to the middle face and all these vertices are distinct. Color the vertex x_f such that the face f will be polychromatic and color the middle face also polychromatic. □

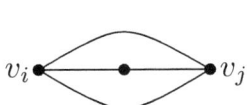

(a) Base graph with 3-faces

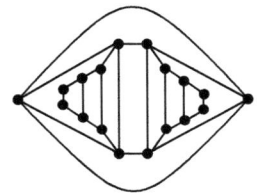

(b) Base graph with 4- and 9-faces

Figure 3.11: Gadgets for the reduction.

Theorem 3.37. *L-PLANE-POLY-3-COLORABILITY*

(i) ... *is in* P *for* $L = \{2, 3\}$.
(ii) ... *is trivial if L contains only even numbers.*
(iii) ... *is* NP-*complete for* $L \supseteq \{3, s\}, s \geq 4$.
(iv) ... *is* NP-*complete for* $L \supseteq \{4, t\}, t \geq 5$ *odd*.
(v) ... *is trivial if* $L \subseteq \{6, \ldots\}$.

Proof. If $g(G) < 3$ then G is certainly not polychromatically 3-colorable. Thus we can assume that $g(G) \geq 3$.

(i) Theorem 3.22 gives a polynomial time checkable criterion for graphs with 3-faces only.

(ii) Because G is bipartite we have $g(G) \geq 4$ and therefore G is polychromatically 3-colorable by Theorem 3.26.

(iii) If $s \geq 5$ is odd then we substitute each edge with a copy of the base graph Figure 3.11(a) but start with a graph which contains only s-faces. By Corollary 3.35(iii) the proper 3-coloring problem restricted to such graphs is NP-hard. Each proper 3-coloring of the s-faces is also a polychromatic 3-coloring and therefore the old graph is properly 3-colorable if and only if the new graph is polychromatically 3-colorable.

If s is even then we start with a graph G with 3- and s-faces only and substitute each edge with a copy of the base graph as in Figure 3.11(a) and extend each s-face as described in Lemma 3.36. Then it holds that the new graph is polychromatically 3-colorable if and only if G is properly 3-colorable.

(iv) Start with a graph G containing only t-faces. We can modify our base graph as indicated in Figure 3.11(b) such that we only have 4-faces and t-faces (and the outer face). By substituting every edge from the input graph G with this new gadget we get G'. The new graph G' has only 4- and t-faces and in every polychromatic 3-coloring of G' the vertices corresponding to the endpoints of edges in G are colored with different colors (Observation 1). Moreover, there exists a 3-coloring of the base graph where v_i, v_j have different colors and all bounded faces are polychromatic. Because t is odd, every proper 3-coloring of G can be extended to a polychromatic 3-coloring of G'. Applying Corollary 3.35(iv) shows the NP-hardness.

(v) Theorem 3.20 implies that all these graphs are polychromatically 3-colorable. □

This result covers all cases except when 5 is the smallest number in L. If $p(5) \geq 3$, which we do not know at the moment, then also $\{5,\ldots\}$-PLANE-POLY-3-COLORABILITY is trivial and the characterization would be complete.

Also note that our base graphs for the Cases (iii), (iv) contain multiple edges and at the moment we do not know whether the results carry over if we restrict to simple graphs.

Finally we prove Theorem 3.29(ii) for $k = 4$.

Proposition 3.38. $\{4\}$-*PLANE-POLY-4-COLORABILITY restricted on simple graphs is* NP-*complete.*

3.5. Complexity Results for Plane Graphs

Proof. Again, we use PLANE-PROPER-3-COLORABILITY. Let G be a simple plane graph. We add a new vertex x_e on each edge $e = \{u, v\} \in E(G)$ and replace the edge $\{u, v\}$ by a path of length two with vertices u, x_e, v. For each face $f \in F(G)$ we add a vertex v_f, place it into the interior of f, and connect v_f to the vertices of f as encountered when traversing the boundary of f in either direction. This yields a new plane simple graph G' where all faces have size exactly 4. See Figure 3.12 for an example.

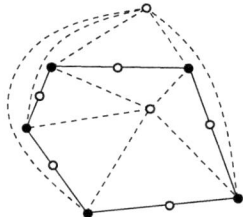

Figure 3.12: Constructing G'.

We claim that G is properly 3-colorable if and only if G' is polychromatically 4-colorable. If G is properly 3-colorable with colors $1, 2$ and 3, then we extend this coloring χ in G' such that each vertex v_f corresponding to a face f in G gets color 4 and the vertex x_e with neighbors u and v gets the color $\{1, 2, 3\} \setminus \{\chi(u), \chi(v)\}$. In this way each face of G' is polychromatic and therefore the whole coloring χ is polychromatic.

Now let us fix a polychromatic 4-coloring χ' of G'. Let v_f be any vertex of G' corresponding to a face f of G. Without loss of generality suppose that v_f has color 4. Then for each edge $e = \{u, v\} \in E(G)$ which is incident to f the vertices $u, x_e, v \in V(G')$ have to get the colors $1, 2$ or 3. Henceforth for every face g of G that shares an edge with f, the vertex v_g gets color 4 as well. Since the dual graph G^* is connected color 4 "propagates" from face to face and $\chi'(v_{f'}) = 4$ for every face f' of G. Also color 4 appears at no other vertex of G'. Now the coloring restricted to the vertices in G uses only three colors and has to be proper because every 4-face f with vertices u, x_e, v, v_f of G' can only be polychromatic if all of its four vertices are colored with distinct colors, and in particular u and v get distinct colors. □

> Ich muess mich nid andersch aleggä,
> wenn ich so redä.
> Ich muess mich nid schträälä,
> wenn ich so redä
> und ich cha mit bluttä Fiässä
> durs heech Gras und under d Lyt,
> wenn ich so redä.
> Muess nid scheen tue,
> wenn ich ebbis gäärä ha.
> Es tuets, wenn ich sägä:
> Ich mag dich wool.
> Und ich traim i dere Schpraach.
>
> Julian Diller

Chapter 4

Extremal Satisfiability

The satisfiability problem was the first problem proven to be NP-complete and therefore it is sometimes also called the "mother" of NP-complete languages. Every problem in the class NP can be reduced to SAT. We will begin this chapter by showing an endcoding of proper k-colorability of graphs into SAT. This will serve us as a illustrating example for the definition of the S-SAT problem, which will be formally introduced in Section 4.1.

Let $k \in \mathbb{N}$ be fixed and $G = (V, E)$ an instance (graph) of the proper k-coloring problem $\text{PROPCOL}(k)$, i.e., we want to decide whether there exists a k-coloring of the vertices V such that there is no monochromatic edge in E. A k-coloring of the vertices assigns to each vertex $v \in V$ *exactly* one of the colors $\{1, 2, \ldots, k\}$. Assume that $V = [n]$. We introduce boolean variables $x_{i,c}$ for $i \in V$ and $c \in [k]$, where $x_{i,c}$ indicates whether the vertex i receives the color c. For $T \subseteq [n] \times [k]$ define

$$\text{AtLeastOneOne}(T) = \bigvee_{t \in T} x_t \qquad (4.1)$$

$$\text{AtMostOneOne}(T) = \bigwedge_{\substack{t,t' \in T, \\ t \neq t'}} \bar{x}_t \vee \bar{x}_{t'} \qquad (4.2)$$

If both formulas hold then exactly one of the variables with index in T is set to true.

$$\text{OneIsOne}(T) = \text{AtLeastOneOne}(T) \wedge \text{AtMostOneOne}(T).$$

Every assignment to the variables $x_{i,j}$ encoding a k-coloring can be expressed as $\bigwedge_{i \in V} \text{OneIsOne}(R_i)$ where $R_i = \{(i,1),(i,2),\ldots,(i,k)\}$. No edge $\{i,j\} \in E$ is monochromatic in a proper k-coloring which can be expressed by

$$\text{ProperEdge}(i,j) = \bigwedge_{c \in [k]} \bar{x}_{i,c} \vee \bar{x}_{j,c} \qquad (4.3)$$

Putting these things together we define the following CNF formula.

$$F(k,G) = \bigwedge_{i \in V} \text{OneIsOne}(R_i) \wedge \bigwedge_{\{i,j\} \in E} \text{ProperEdge}(i,j).$$

Note that the formula can be constructed in polynomial time. For different graphs on the same vertex set $V = [n]$ the first part will always be the same and therefore the essential information is in the second part. If we restrict the SAT-problem such that only *some* assignments of $\{0,1\}^*$ are allowed, then we can capture this better. Let $N = nk$ and define

$$S_N = \{(x_{i,j})_{i \in [n], j \in [k]} \in \{0,1\}^N : \text{OneIsOne}(R_i), \text{ for all } i \in [n]\}.$$

There exists a proper k-coloring of G if and only if the formula

$$\bigwedge_{\{i,j\} \in E} \text{ProperEdge}(i,j)$$

is satisfiable with an assignment from S_N. For $k = 3$ we have $S_{3n} = (001|010|100)^n$ and because it is well-known that proper 3-colorability is NP-hard, the satisfiability problem restricted to such assignments is NP-hard as well.

This phenomena occurrs quite often in encodings of problems in NP, i.e., there is one part of the formula which is the same for all instances of the same size. Therefore, it is preferable to split this part from the remaining part which captures the instance-specific information. This is exactly how we define the S-SAT problem.

4.1 Problem Description

The S-SAT problem is a variant of the SAT problem where we allow only *some* assignments to be considered. For simplicity of notation, we agree that the boolean variables are named $v_1, \ldots, v_n$, and they are ordered like this. Since we assume that the set of variables is ordered, we can interpret $x \in \{0,1\}^n$ as a truth assignment of the variables $v_1, v_2, \ldots, v_n$.

> **Given:** $S \subseteq \{0,1\}^*$
> **Input:** formula F, ordered variable set $V \supseteq \mathrm{vbl}(F)$
> **Output:** Yes, if there exists an assignment $\alpha \in \{0,1\}^{|V|} \cap S$ that satisfies F, otherwise no.

We define $S_n := S \cap \{0,1\}^n$ for all $n \in \mathbb{N}$ and call these sets the *levels* of S. Note that V is part of the input but we do not require every variable in V to occur in F. This is the same as to say that $f(x, y, z) = x$ is a function in three variables. The *complexity* of the input is the size of the formula plus the size of the variable set. This refers to the time needed to evaluate the formula F for an assignment $x \in \{0,1\}^V$ (up to some polynomial factors in n): we need $|V|$ time to read an assignment and the size of the formula captures the time to evaluate F. We want to point out that S is fixed and not part of the input.

Example 1. For $S = \{0,1\}^*$, the S-SAT problem is the normal SAT problem.

Example 2. For $S = (001|010|100)^*$, the S-SAT problem contains all instances of the proper 3-colorability problem. Thus, this S-SAT problem is NP-hard, even restricted to 2-CNF formulas.

Example 3. Let $k \in \mathbb{N}$ be fixed and $G = (V, E)$ a plane multigraph with faces F. For $f \in F$ we denote by $V(f)$ the vertices of this face. Instead of (4.3) we define

$$\text{Polychromatic}(f) := \bigwedge_{c \in [k]} \bigvee_{i \in V(f)} x_{i,c}. \qquad (4.4)$$

By Theorem 3.29 we know that polychromatic k-colorability is NP-hard for $k = 3, 4$. Thus, the S-SAT problem with $S = (001|010|100)^*$ is NP-hard even restricted to $*$-planar CNF formulas where each variable occurs only positive.

Example 4. Let H be fixed and $G = (V, E)$ an instance. An edge-coloring using k colors can be encoded similarly as before by using the formula $\text{OneIsOne}(T)$. The property that the edge set Z of a subgraph of G isomorphic to H is not monochromatic, can be encoded by

$$\text{NotMonochromatic}(Z) := \bigwedge_{c \in [k]} \bigvee_{t \in Z} \bar{x}_{t,c} \qquad (4.5)$$

A special case is that $H = K_{1,2}$, where a graph G is not H-Ramsey with k colors if and only if G is proper k-edge colorable. Since this problem is NP-hard, Theorem 3.11, the property of being H-Ramsey is co-NP-hard.

Example 5. Let $G = (V, E)$ be a graph with $V = \{1, 2, \ldots, n\} = [n]$. A Hamiltonian cycle in G is a permutation of the vertices such

4.1. Problem Description

that between consecutive vertices there is an edge. A permutation is a bijection $\pi : [n] \to [n]$. For $i \in [n], j \in [n]$, let $x_{i,j}$ be a boolean variable which is true if and only if $\pi(i) = j$. Let $R_i = \{(i,1),(i,2),\ldots,(i,n)\}$ and $C_j = \{(1,j),(2,j),\ldots,(n,j)\}$ (one can imagine these index sets as rows and columns of the $(n \times n)$-array). Then an assignment to the $x_{i,j}$'s encode a permutation if and only if the following condition is satisfied.

$$\bigwedge_{i \in [n]} \text{OneIsOne}(R_i) \wedge \bigwedge_{j \in [n]} \text{OneIsOne}(C_j) \tag{4.6}$$

Moreover, they correspond to a Hamiltonian cycle in G if between consecutive elements $\pi(i), \pi(i+1)$ there is an edge, i.e., there are no non-edges between consecutive elements. By identifying $n+1$ with 1, we can encode this by the following formula.

$$\bigwedge_{\{i,j\} \notin E} \bigwedge_{k \in [n]} \bar{x}_{i,k} \vee \bar{x}_{j,k+1} \tag{4.7}$$

The formulas (4.6) and (4.7) together are a polynomial encoding of the Hamiltonian cycle problem into SAT. Furthermore, set $N = n^2$ and define

$$S_N = \{(x_{i,j})_{i \in [n], j \in [k]} \in \{0,1\}^N : (4.6) \text{ holds }\}.$$

There exists a Hamiltonian cycle in G if there exists an assignment in S_N that satisfy formula (4.7). Since the Hamiltonian cycle problem is NP-hard, also this S-SAT problem is NP-hard.

Outlook. A family S is called *asymptotically exponential* if $|S_n| \in \Omega(\alpha^n)$ for some $\alpha > 1$. Cooper [27] asked whether for all asymptotically exponential languages S, the S-SAT problem is NP-hard. We will answer this question negatively in Section 4.7. This gives rise to the following two questions.

(1) For which languages S is the S-SAT problem NP-hard?
(2) For which languages S is the S-SAT problem in P?

In Section 4.4 we show that it is unlikely that for an asymptotically exponential family S the S-SAT problem is in P. In Section 4.5, we show that for context-free languages S the S-SAT problem is in P, if $|S_n|$ is polynomial in n, and it is NP-hard otherwise.

4.2 Some Observations

If there are only a few elements in S, then the S-SAT problem cannot be very hard. To make this more precise we state

Proposition 4.1. *If $|S_n|$ is polynomial in n and S_n can be enumerated in polynomial time then S-SAT is in* P.

If we view S itself as a language over the alphabet $\{0,1\}$, and therefore as a decision problem, we get the following connection:

Proposition 4.2. *S can be reduced to S-SAT in polynomial time.*

Proof. Given some $x = (x_1, \ldots, x_n) \in \{0,1\}^n$ and define the 1-CNF formula
$$F_x := \bigwedge_{i: x_i = 1} v_i \wedge \bigwedge_{i: x_i = 0} \bar{v}_i.$$
Then x is the unique assignment in $\{0,1\}^n$ that satisfies the formula F_x. Hence, F_x is S-satisfiable if and only if $x \in S_n$. Clearly, this is a polynomial reduction from S to S-SAT. $\square$

Corollary 4.3. *If S as decision problem is* NP-*hard, then S-SAT is* NP-*hard, even restricted to 1-CNF formulas.*

The above considerations are showing that S-SAT is difficult for some S. We continue by proving that S-SAT is difficult for every asymptotically exponential S. More precisely, we demonstrates how we can employ a fast S-SAT algorithm, if existent, to solve SAT in significantly less than 2^n steps. We write $O^*(f(n))$ if we neglect polynomial factors.

4.2. Some Observations

Proposition 4.4. *Suppose there is some S with $|S_n| \in \Omega(\alpha^n)$ for $1 < \alpha < 2$. If S-SAT can be decided in time $O^*(\beta^n)$, then there is a randomized Monte Carlo algorithm for SAT with running time $O^*((2\beta/\alpha)^n)$.*

Proof. Let F be a formula over a set V of variables, and let x be an assignment. For each variable $v \in V$, *switch* v with probability $1/2$, i.e., invert all its occurrences in F and its value according to the assignment x, resulting in a new formula F' and a new assignment x'. The assignment x' satisfies F' if and only if x satisfies the original formula F. Moreover, x' is uniformly distributed over $\{0,1\}^n$. First, assume that F is satisfiable, then the formula F' is S-satisfiable with probability $\Pr[x' \in S_n] \geq (\alpha/2)^n$. This can be tested in time $O^*(\beta^n)$. After repeating this process $(2/\alpha)^n$ times, the probability that at least one of the randomly generated formulas is S-satisfiable, is at least $1 - 1/e$, hence constant. On the other hand, if F is unsatisfiable, it will not become satisfiable by switching variables. We therefore have a Monte Carlo algorithm with running time $(2/\alpha)^n O^*(\beta^n)$. □

There are no known algorithms for SAT running in time $O^*(\gamma^n)$ for $\gamma < 2$, not even randomized ones. Proposition 4.4 with $\beta < \alpha$, therefore, is a first indication that S-SAT is a difficult problem.

Example. The currently best known deterministic algorithms for 3-SAT [79, 14] are based on the algorithm of Dantsin et al. [28]. In fact, it can be viewed as a derandomized version of the randomized algorithm in the proof of Proposition 4.4: Let the *Hamming distance* $d(x,y)$ of two vectors $x, y \in \{0,1\}^n$ be the number of coordinates in which they differ. The *Hamming Ball* of radius r around x is the set $B_r(x) := \{y \in \{0,1\}^n \mid d(x,y) \leq r\}$. We look at the family $S_n = B_{\rho n}(\mathbf{0})$ where $0 < \rho < 1$ is some constant. Then

$$|S_n| = \sum_{i=0}^{\lfloor \rho n \rfloor} \binom{n}{i} \approx 2^{H(\rho)n}, \quad H(t) = -t \log t - (1-t) \log(1-t).$$

Thus, $S = (S_n)_{n \geq 0}$ is an asymptotically exponential family. For 3-CNF formulas, S-SAT can be decided in $O^*(3^{\rho n})$ steps (by splitting on 3-clauses), which for appropriately chosen ρ is much smaller than $2^{H(\rho)n}$. By choosing many Hamming balls centered at different points randomly and by choosing the optimal value of ρ this yields an algorithm deciding 3-SAT in $O^*(1.5^n)$ steps. Note that choosing a random point as center of the Hamming ball is equivalent to switching the formula randomly and keeping the Hamming ball centered at $(0, \ldots, 0)$ all the time. It takes some additional effort to derandomize the algorithm, see [28].

4.3 S-SAT and the VC-dimension

To obtain a systematic way of proving NP-hardness of S-SAT (if possible), we use the notion of *shattering* and the *Vapnik-Chervonenkis-dimension* (VC-dimension). These concepts were first introduced by Vapnik and Chervonenkis [89]. Let V be the set of variables containing $v_1, v_2, \ldots, v_n$. We say $I \subseteq [n]$ is *shattered* by S_n if any assignment to $V_I := \{v_i \mid i \in I\}$ can be realized by S_n. Formally, for every $x \in \{0,1\}^{|I|}$ there is a $y \in S_n$ with $y|_I = x$, where $y|_I$ denotes the $|I|$-bit vector $(y_i)_{i \in I}$. The *VC-dimension* d_{VC} is the size of a largest shattered set. Obviously, $0 \leq d_{\text{VC}}(S_n) \leq n$. The intuition is that large sets have large VC-dimensions. This is quantified by the following lemma, which was proven several times independently, see for example [77, 83, 89].

Lemma 4.5. *Suppose $d_{\text{VC}}(S_n) \leq d \leq n/2$. Then*

$$|S_n| \leq \sum_{i=0}^{d} \binom{n}{i} \leq 2^{H(\frac{d}{n})n}$$

where $H(x) = -x \log(x) - (1-x) \log(1-x)$ is the binary entropy function.

We will give here a non-standard proof of this lemma using satisfiability which shows also an interesting fact about the number of satisfying assignments for a CNF formula.

4.3. S-SAT and the VC-dimension

Proof. By assertion $d_{\text{VC}}(S_n) \leq d$, hence no index set $I \subseteq [n]$ of size $(d+1)$ can be shattered by S_n. This means that for every $I \subseteq [n]$ with $|I| = d+1$ there is some $x(I) \in \{0,1\}^{d+1}$ such that all $y \in S_n$ satisfy $y|_I \neq x(I)$. For $I = \{i_1, \ldots, i_{d+1}\} \subseteq [n]$ consider the disjunction $C_I := v_{i_1}^{1-x(I)_1} \vee \ldots \vee v_{i_{d+1}}^{1-x(I)_{d+1}}$, where $v^1 := v$ and $v^0 := \bar{v}$. Every assignment *not* satisfying C_I must agree with $x(I)$ in the variables with indices in I. Thus all elements of S_n satisfy C_I. Define

$$F = \bigwedge_{I \subseteq [n], |I| = d+1} C_I,$$

which is a $(d+1)$-CNF and $S_n \subseteq \text{sat}(F)$, where $\text{sat}(F)$ denotes the set of satisfying assignments for F. Let F' be the unsigned version of F, i.e., replace every negative literal by its positive counterpart. This F' consists of all unsigned $(d+1)$-clauses over n variables, and $x \in \{0,1\}^n$ satisfies F' iff x contains at most d many 0's. Therefore $\text{sat}(F') = B_d(\mathbf{1})$ and $|\text{sat}(F')| = \text{vol}(n, d) \leq 2^{H(d/n)n}$ where the last inequality for $0 \leq d \leq 1/2$ is well known. It is enough to show that $|\text{sat}(F)| \leq |\text{sat}(F')|$ which will be done in the next lemma. □

Lemma 4.6. *Let F be any CNF formula and let F' be the unsigned version of F. Then $|\text{sat}(F)| \leq |\text{sat}(F')|$.*

Proof. Let v be a variable in F which occurs also negatively. Define F_v to be the CNF formula obtained from F by replacing every occurrence of the literal $\bar{v}$ by v, i.e., v occurs only as positive literals in F_v. Let x be an assignment and x' the assignment with $x'(v) = 1$ and $x'(w) = x(w)$ for $w \neq v$. If x satisfies F then x' satisfies F_v. Thus we have a function from $\text{sat}(F)$ to $\text{sat}(F_v)$ which we can make injective: If we obtain x' twice then x' and the assignment agreeing in x except setting v to 0 are satisfying assignments for F and therefore also for F_v. Thus $|\text{sat}(F)| \leq |\text{sat}(F_v)|$. By repeatedly applying this procedure for every variable leads to the unsigned version F' of F and the inequality follows. □

By Lemma 4.5 every asymptotically exponential family S fulfills $d_{\text{VC}}(S_n) \in \Omega(n)$. Our goal is to use this to show that S-SAT is then

NP-hard or unlikely to be in P for an asymptotically exponential families S. Such results become more general if we enlarge the class of families S for which they apply. Instead of using the notion of asymptotically exponential we use a generalization of it. As we have seen in the encodings of NP-hard problems to S-SAT there might be some levels S_n which do not contain any elements.

Definition 4.7. *A monotone increasing sequence $Q = (n_j)_{j \in \mathbb{N}} \subseteq \mathbb{N}$ has polynomial gaps if there is a polynomial $p(n)$ such that*

$$n_{j+1} \leq p(n_j)$$

for all $j \in \mathbb{N}$.

A sequence (n_j) can increase exponentially in j and still have polynomial gaps. For example, define $n_j = 2^j$. Then $n_{j+1} = 2n_j$, so $p(n) = 2n$ shows that this sequence has polynomial gaps. Note that we can always assume without loss of generality that $p(n)$ is strictly increasing.

Definition 4.8. *The family $(S_n)_{n \geq 1}$ is called* exponential *if there exists $\alpha > 1$ and a sequence Q with polynomial gaps such that*

$$\forall n \in Q : |S_n| \geq \alpha^n .$$

We also say $S = \bigcup_{n \geq 1} S_n$ has exponential size.

For example, families with $|S_n| \in \Omega(\alpha^n)$ are exponential (but we additionally allow to have some "gaps"). It is easy to see that the encodings from the beginning of this chapter for k-colorings are an exponential family but for permutations they are not. For permutation we have $N = n^2$ and $|S_N| = n! \leq 2^{n \log n} = 2^{0.5\sqrt{N} \log N}$ which is subexpontial.

The connection between exponential families and large shattered index set is given trough Lemma 4.5.

Corollary 4.9. *Suppose $S \subseteq \{0,1\}^*$ is exponential. Then there is a polynomial $q(n)$ such that for each $n \in \mathbb{N}$ there exists $N \leq q(n)$ and an index set $I \subseteq [N]$ with $|I| \geq n$ such that I is shattered by S_N.*

4.3. S-SAT and the VC-dimension

Proof. Let $(n_j)_{j \in \mathbb{N}}$ be the sequence with polynomial gaps corresponding to the exponential family S, i.e., there is an $\alpha > 1$ and a polynomial p such that $n_{j+1} \leq p(n_j)$ and $|S_{n_j}| \geq \alpha^{n_j}$ for all j. Let $\delta \in (0, 1/2]$ such that $H(\delta) = \log \alpha$ and define the polynomial q by $q(n) = p(n/\delta)$. Choose k such that $n_k \leq \frac{n}{\delta} \leq n_{k+1} =: N$. By Lemma 4.5, $d_{\mathrm{VC}}(S_N) \geq \delta N \geq n$, so there exists a shattered set $I \subseteq [N]$ with $|I| \geq n$. Note that $N = n_{k+1} \leq p(n_k) \leq p(n/\delta) = q(n)$, as required. $\square$

Although we know that a large shattered set exists, it is not clear how we can compute it efficiently. Let us for the moment assume that we can. Then there is a polynomial reduction from SAT to S-SAT:

Theorem 4.10. *Let $S \subseteq \{0,1\}^*$ be of exponential size and let $p(n)$ be a polynomial. Suppose that for all n, we can compute, in time polynomial in n, some number $N \leq p(n)$ and some index set $I \subseteq [N]$ with $|I| \geq n$ that is shattered by S_N. Then S-SAT is NP-hard.*

Proof. Let F be a formula over the variables $V_n = \{v_1, \ldots, v_n\}$. We construct a new formula F' over V_N by renaming each v_j occurring in F into v_{i_j} where $I \supseteq \{i_1, \ldots, i_n\}$. We claim that F is satisfiable iff F' is S-satisfiable. Suppose $x \in \{0,1\}^n$ satisfies F. Because I is shattered by S_N there is an assignment $y \in S_N$ that agrees with x in the variables $(v_{i_1}, \ldots, v_{i_n})$ and thus F' is S-satisfiable. The reverse direction is clear. This polynomial reduction shows that S-SAT is NP-hard. $\square$

Why does this method not work in general? The difficulty is that we do not know which subset of variables is shattered, we only know that there is one. The result of Papadimitriou and Yannakakis [73] states that computing the VC-dimension of an explicitly given S_n (of size not necessarily exponential in n) is LOGNP-complete, hence unlikely to be in P. If S is exponential then a brute force approach can be made to compute the VC-dimension in time polynomial in $|S_n|$. However, if computing the VC-dimension takes time polynomial in $|S_n|$, then this is not useful because $|S_n|$ itself is exponential in n.

Example. We give an example of S where the VC-dimension is uncomputable, and still there is a straightforward reduction from SAT to S-SAT. Let $U \subseteq \mathbb{N}$ be an undecidable set, e.g.

$$U := \{n \in \mathbb{N} \mid T_n \text{ halts on empty tape}\},$$

where T_n is the nth Turing machine in some sensible enumeration. Then define

$$L_n = \{w0^{\lceil 2n/3 \rceil} \mid w \in \{0,1\}^{\lfloor n/3 \rfloor}\},$$
$$R_n = \{0^{\lceil n/2 \rceil} w \mid w \in \{0,1\}^{\lfloor n/2 \rfloor}\}.$$

Finally, set

$$S_n = \begin{cases} L_n, & \text{if } n \in U; \\ R_n, & \text{otherwise.} \end{cases}$$

Clearly, the VC-dimension of S_n is either $\lfloor n/3 \rfloor$ or $\lfloor n/2 \rfloor$, but is it undecidable which holds. Still, there is a simple reduction from SAT to S-SAT. For a formula F with n variables $v_1, \ldots, v_n$, choose $N = 3n$. Then in S_N, we know that either $\{1, \ldots, N/3\}$ or $\{2N/3, \ldots, N\}$ is shattered. Let F' be the same formula as F, but with each variable v_i renamed into v_{N+1-i}. Certainly, if $N \in U$, then F is S-satisfiable, and if $N \notin U$, then F' is S-satisfiable. So $\varphi(F) := F \vee F'$ is S-satisfiable if and only if F is satisfiable. Hence φ is a polynomial time reduction.

4.4 S-SAT and Polynomial Circuits

We will prove a result that is "almost as good" as proving NP-completeness: if S-SAT is in P for some exponential S, then SAT has polynomial circuits. We will briefly introduce the notations used from circuit theory (see [90, 72] for a more elaborate discussion).

Let $B = \{\neg, \wedge, \vee\}$ be the basis. A *boolean circuit* over n variables $x_1, \ldots, x_n$ is a directed acyclic graph $G = (V, E)$ with labels at each vertex fulfilling the following properties.

4.4. S-SAT and Polynomial Circuits

(i) If $v \in V$ has in-degree 0, then it is labeled with one of the variables $x_1, \ldots, x_n$ (these vertices are called input gates).
(ii) If $v \in V$ has in-degree 1, then it is labeled with $\neg$.
(iii) If $v \in V$ has in-degree 2, then it is labeled either with $\wedge$ or $\vee$.
(iv) There is exactly one vertex $v \in V$ with out-degree 0 (this vertex is called the output gate).

Given specific values for the variables $x_1, \ldots, x_n$ we can compute the value of each inner node (inductively) in the obvious way. The *size* of a boolean circuit is the number of gates and the *depth* is the length of the longest directed path in G.

A *circuit family* is a sequence $\mathcal{C} = (C_1, C_2, \ldots)$ of boolean circuits, where each C_n has n input gates. If each C_n has exactly one output gate, then $\mathcal{C}$ computes a function $f : \{0,1\}^* \to \{0,1\}$, or equivalently, decides a language $L \subseteq \{0,1\}^*$. If the size of C_n grows polynomially in n, then $\mathcal{C}$ is a *polynomial circuit family*.

Definition 4.11. *A language $L \in \{0,1\}^*$ has* polynomial circuits *if there exist a polynomial circuit family $(C_i)_{i \in \mathbb{N}}$ that decides L. The class of all languages L with polynomial circuits is denoted by* P/poly.

Let L be a language which has polynomial circuits $(C_i)_{i \in \mathbb{N}}$. If there exists an algorithm that computes C_n in time polynomial in n, then clearly $L \in \mathsf{P}$. The other direction holds as well. We also say here that L has *uniform* polynomial circuits. There are undecidable languages L with nonuniform polynomial circuits, e.g. unary languages defined by an undecidable problem.

Theorem 4.12. *If S-SAT is in* P *for some exponential S, then* SAT *has (possibly nonuniform) polynomial circuits.*

Proof. From Corollary 4.9, we know that for each n there exists an $N \leq q(n)$ and an index set $I \subseteq [N]$ with $|I| \geq n$ such that I is shattered by S_N. For each n, there is a boolean circuit of polynomial size that takes a formula F over n variables as input and outputs a formula F'

over N variables, where F' is identical to F, but with all variables from F replaced by variables in I. Note that the circuit *exists*, though it might not be constructible in polynomial time. By assumption, there is a second circuit of polynomial size deciding S-SAT for formulas with N variables. Combining these two circuits yields a polynomial circuit deciding SAT. □

If SAT has polynomial circuits then all problems in NP have polynomial circuits, because SAT is NP-complete. There are strong reasons to believe that this is not the case.

Theorem 4.13 (Karp and Lipton [56]). *If* $\mathsf{NP} \subseteq \mathsf{P}/\mathsf{poly}$ *then the polynomial hierarchy collapses to its second level, i.e.,* $\mathsf{PH} = \Sigma_2 \mathsf{P}$.

It might be possible that NP has polynomial circuits and still $\mathsf{P} \neq \mathsf{NP}$. On the other hand if $\mathsf{P} = \mathsf{NP}$ then also $\Sigma_2 \mathsf{P} = PH$. Hence, this result is weaker than proving NP-hardness for S-SAT in general. However, there has not been shown any collapse between some levels in the polynomial hierarchy for the last 27 years.

There are some improvements on the Karp-Lipton Theorem which we want to mention here (for the involved complexity classes please visit the Complexity Zoo [1]):

(i) If $\mathrm{SAT} \in \mathsf{P}/\mathsf{poly}$ then $\mathrm{PH} = \mathsf{ZPP}^{\mathsf{NP}}$ [60].
(ii) If $\mathrm{SAT} \in \mathsf{P}/\mathsf{poly}$ then $\mathrm{PH} = \mathsf{S}_2^\mathsf{P}$ [22].
(iii) If $\mathrm{SAT} \in \mathsf{P}/\mathsf{poly}$ then $\mathrm{MA} = \mathrm{AM}$ [7].

4.5 S-SAT for Context-Free Languages S

Ginsburg and Spanier [43] showed that every context-free language S is either polynomial or asymptotically exponential. In this section, we prove that S-SAT is NP-complete if S is an exponential, context-free language and S-SAT is in P if S is a polynomial, context-free language.

In the following, we denote the nonterminal symbols appearing in a context-free grammar for S by upper case letters S_0, A, B, C, and so on,

4.5. S-SAT for Context-Free Languages S

where S_0 is the starting symbol. The only terminal symbols are $0, 1$. All rules in a context-free grammar are of the form $A \Rightarrow w$ for a word w possibly containing nonterminals. $A \stackrel{*}{\Rightarrow} w$ means that w can be derived from A in finitely many steps. Finally, the length of a word $x \in \{0,1\}^*$ is denoted by $|x|$.

Example. Let $S = (100|010|001)^*$ be the family corresponding to 3-colorings of vertices. The following substitution rules with start symbol B_3 generate S.

$$A_{i-1} \to 0A_i | 1B_i, \qquad i = 1, 2, 3$$
$$B_{i-1} \to 0B_i, \qquad i = 1, 2, 3$$
$$B_3 \to \varepsilon | A_0$$

This grammar is context-free, actually it is even regular.

Let S be a context-free, exponential language which is generated by the grammar G. All calculations on the grammar can be done in advance and therefore do not contribute to the running time. In particular, we may assume that G does not contain *useless* nor *unreachable* nonterminal symbols, i.e., for every nonterminal A, we have $A \stackrel{*}{\Rightarrow} x$ for some $x \in \{0,1\}^*$, and $S \stackrel{*}{\Rightarrow} w$ for some w with $A \in w$. We call such a grammar *reduced*.

Theorem 4.14. *Let $S \subseteq \{0,1\}^*$ be a context-free polynomial language given by a context-free grammar G. Then S-SAT is in* P.

Proof. By the discussion above it is enough to prove the lemma for reduced context-free grammars G. Moreover, we can assume that all substitution rules are of the form $A \to UV$ or $A \to \varepsilon$ (Chomsky normal form). Define for each nonterminal A, $W_A^0 = \{\varepsilon\}$ if $A \to \varepsilon$ and for $1 \leq k \leq n$

$$W_A^k := \{w \in \{0,1\}^k : A \Rightarrow^* w\}.$$

It is easy to see that, since S is polynomial, all these sets are polynomially bounded in n. They can be computed inductively by

$$W_A^k = \bigcup_{A \to UV} \bigcup_{0 \leq \ell \leq k} W_U^\ell \times W_V^{k-\ell}.$$

This leads to a polynomial enumeration of $W_{S_0}^n = S \cap \{0,1\}^n$. By Proposition 4.1 the S-SAT problem is then in P. □

For a nonterminal A, define

$$\ell(A) := \left\{ x \in \{0,1\}^* \mid \exists y \in \{0,1\}^* : A \overset{*}{\Rightarrow} xAy \right\},$$
$$r(A) := \left\{ y \in \{0,1\}^* \mid \exists x \in \{0,1\}^* : A \overset{*}{\Rightarrow} xAy \right\}.$$

Call some $X \subseteq \{0,1\}^*$ *commutative* if $x_1 x_2 = x_2 x_1$ for all $x_1, x_2 \in X$.

Theorem 4.15 (Ginsburg [43, Theorem 5.5.1])**.** *Let G be a reduced context-free grammar and let $L(G)$ be the language generated by G. Then $|L(G) \cap \{0,1\}^n|$ is polynomial in n if and only if for every nonterminal A, $\ell(A)$ and $r(A)$ are commutative.*

It is not hard to prove that if there is a nonterminal A such that $\ell(A)$ (or $r(A)$) is commutative then the language is asymptotically exponential. Actually, we will perform a similar argument for proving the next theorem.

Theorem 4.16. *Suppose $S \subseteq \{0,1\}^*$ has exponential size and is a context-free language. Then S-SAT is NP-complete.*

Proof. We will show how to compute large shattered sets for every n. Let G be a reduced context-free grammar for S. Since S has exponential size, $|S_n|$ is surely not polynomial in n. Therefore, Theorem 4.15 implies that there is a nonterminal A such that $\ell(A)$ or $r(A)$ is not commutative. Since we have only to prove that there *exists* a polynomial reduction from SAT to S-SAT, the *existence* of such a nonterminal is enough. Suppose without loss of generality that $\ell(A)$ is not commutative, and

4.5. S-SAT for Context-Free Languages S

let $x_1, x_2 \in \ell(A)$ such that $x_1 x_2 \neq x_2 x_1$. Hence, there is a position i such that without loss of generality $(x_1 x_2)_i = 0$ and $(x_2 x_1)_i = 1$. By definition, there are $y_1, y_2 \in \{0,1\}^*$ such that $A \stackrel{*}{\Rightarrow} x_1 A y_1$ and $A \stackrel{*}{\Rightarrow} x_2 A y_2$. By applying k times either $A \stackrel{*}{\Rightarrow} x_1 x_2 A y_2 y_1$ or $A \stackrel{*}{\Rightarrow} x_2 x_1 A y_1 y_2$, we can create arbitrary 0s and 1s at the positions $i + k \cdot |x_1 x_2|$ for any k. In order to reach A from S_0, we use $S_0 \stackrel{*}{\Rightarrow} aAb$, and in the end we use $A \stackrel{*}{\Rightarrow} w$ to obtain a word in $\{0,1\}^*$ for some $a, b, w \subseteq \{0,1\}^*$. Hence if we set $N := |a| + |b| + |w| + n(|x_1 x_2| + |y_1 y_2|)$, then $I := \{|a| + k|x_1 x_2| + i : 0 \leq k \leq n - 1\}$ is of size n, and it is shattered by S_N. All these calculations can be done in time $O(n)$ and N is linear in n. Thus, by Theorem 4.10, S-SAT is NP-hard. It is clear that S-SAT is in NP if S is context-free, because deciding whether $x \in S$ and verifying that x is satisfying can be done in polynomial time. Therefore, S-SAT is NP-complete. □

Continuation of the example. Let $S = (001|010|100)^*$ be the exponential, context-free language from before. Clearly, $I \subseteq \{1, 4, 7, 10, \ldots\}$ is an index set which is shattered by S_N for $N = |I|$. Look for example at the following SAT-instance

$$F = (v_1 \vee \bar{v}_2) \wedge (v_1 \vee v_2 \vee \bar{v}_3), \quad V = \{v_1, v_2, v_3\}.$$

Then we can encode it as a S-SAT instance

$$F' = (v_1 \vee \bar{v}_4) \wedge (v_1 \vee v_4 \vee \bar{v}_7), \quad V' = \{v_1, v_2, \ldots, v_9\}.$$

F is satisfiable if and only if F' is S-satisfiable. We see that a CNF formula will be mapped to a CNF formula again. One could argue that this is not a "nice" CNF formula, because $v_2, v_3, v_5, v_6, v_8, v_9$ are variables which do not occur in F'. By expanding the index set by 1 and adding a dummy clause one can overcome this drawback.

$$F'' = F' \wedge (v_2 \vee v_3 \vee v_5 \vee v_6 \vee v_8 \vee v_9 \vee v_{10} \vee v_{11} \vee v_{12}),$$
$$V'' = \{v_1, v_2, \ldots, v_{12}\}.$$

For every assignment of the variables v_1, v_4, v_7 there is an assignment in S_{12} that agrees on these variables and additionally sets v_{10} to true. Every variable in F'' appears at least once. Since F is satisfiable, the pair (F'', V'') is S-satisfiable.

4.6 VC-Dimension of Regular Languages

Assume that S is not only context-free but also a *regular* language, and is of exponential size. We will show that a *maximum size* index set I that is shattered by S_n can be computed efficiently, i.e., in time polynomial in n. Compare this to the result from Section 4.5: there, we constructed *fairly large* shattered index sets for values n of our choice but not necessarily one of maximum size.

First, we give a polynomial algorithm to decide whether a given index set I is shattered by S_n. The idea to continue then is to run this algorithm in parallel for all I in such a way that the whole computation remains polynomial. Therefore, we will have to identify different index sets according to their shattering properties.

We know that S is regular, so we have a deterministic finite state machine (DFSM) deciding S. This DFSM has a set $Q = \{q_0, q_1, \ldots, q_{d-1}\}$ of states and a set $A \subseteq Q$ of accepting states, and a start state q_0. It is equipped with a state transition function $\delta : Q \times \{0,1\} \to Q$. This δ can be extended to a function $\hat\delta : \{0,1\}^* \to Q$, where $\hat\delta(w) = q$ if the DSFM starting in q_0 will be in state q after processing the input word w.

For $I \subseteq [n]$, $x \in \{0,1\}^I$ and $y \in \{0,1\}^{[n]\setminus I}$, let $x \circ y$ denote the vector $w \in \{0,1\}^n$ with $w|_I = x$ and $w|_{[n]\setminus I} = y$. The condition that I is shattered by S_n can now be stated as following

$$\forall x \in \{0,1\}^I \; \exists y \in \{0,1\}^{[n]\setminus I} : \hat\delta(x \circ y) \text{ is an accepting state}.$$

We will interpret the states q_i as boolean variables, which are set to true

4.6. VC-Dimension of Regular Languages

for all accepting states and false to the other states. Thus the accepting states A determine a true-false assignment φ_A of the q_i's and for the formula

$$F_n^I := \bigwedge_{x \in \{0,1\}^I} \bigvee_{y \in \{0,1\}^{[n]\setminus I}} \hat{\delta}(x \circ y),$$

we have that φ_A satisfies F_n^I if and only if I is shattered by S_n.

Lemma 4.17. *Let $S \subseteq \{0,1\}^*$ be a regular language given by a deterministic finite state machine as above and $I \subseteq [n]$. Then there is an algorithm deciding whether I is shattered by S_n which runs in time $O(n)$.*

Proof. The number of variables in F_n^I is exactly $|Q| = d$ which is a constant. Moreover the formula is monotone, i.e., every variable occurs only positively. Thus there are at most 2^d different clauses each of size at most d which implies that the size of the formula is bounded by $d2^d$. We will build the formula F_n^I iteratively and in each step we take care that in each clause there are no repeating literals and there are no repeating clauses. Define

$$F_k^I := \bigwedge_{x \in \{0,1\}^I} \bigvee_{y \in \{0,1\}^{[n]\setminus I}} \hat{\delta}(x \circ y|_{\{1,2,\ldots,k\}}).$$

Then $F_0^I = q_0$ and for $k \geq 1$

$$F_k^I = \begin{cases} \bigwedge_{C \in F_{k-1}^I} \bigvee_{q \in C, t \in \{0,1\}} \hat{\delta}(q,t), & \text{if } k \notin I; \\ \bigwedge_{C \in F_{k-1}^I, t \in \{0,1\}} \bigvee_{q \in C} \hat{\delta}(q,t), & \text{if } k \in I, \end{cases} \quad (4.8)$$

where by a slightly abuse of notation the AND goes over all clauses C in F_{k-1}^I and the OR goes over all variables q in C. The time to compute this formula is linear in the size of F_{k-1}^I which is inductively bounded by $d2^d$. We can delete repeating literals in a clause and repeating clause such that the formula F_k^I has size at most $d2^d$ again. Thus the time in each iteration step is constant with respect to n and the overall time for constructing the F_n^I is in $O(n)$. Evaluating F_n^I for the assignment

φ_A given by all the accepting states can be performed in constant time. Altogether we can decide whether I is shattered by S_n in time $O(n)$. □

Theorem 4.18. *If $S \subseteq \{0,1\}^*$ is a regular language given by a DFSM then $d_{\mathrm{VC}}(S_n)$ and a shattered set $I \subseteq [n]$ of this size can be computed in $O(n)$ time.*

Proof. We say that an index set $I \subseteq [k]$ is *k-maximum* if it is of maximum size over all index sets $I' \subseteq [k]$ with $F_k^I = F_k^{I'}$. Define

$$\mathcal{C}_k := \{(F_k^I, |I|) : I \subseteq [k] \text{ is } k\text{-maximum}\}.$$

Clearly, $d_{\mathrm{VC}}(S_n)$ is the largest $|I|$ in $\mathcal{C}_n$ for which the corresponding formula F_k^I is true under φ_A. We have $\mathcal{C}_0 = \{(F_0^\emptyset, 0)\} = \{(q_0, 0)\}$. There are at most 2^{2^d} different formulas over d variables, i.e. $|\mathcal{C}_k| \leq 2^{2^d}$. Let $k \geq 1$ and $(F, |I|) \in \mathcal{C}_{k-1}$. Define in analogy to (4.8):

$$\mathrm{ext}((F, |I|), b) := \begin{cases} (\bigwedge_{C \in F} \bigvee_{q \in C, t \in \{0,1\}} \hat{\delta}(q, t), |I|), & \text{if } b = 0; \\ (\bigwedge_{C \in F, t \in \{0,1\}} \bigvee_{q \in C} \hat{\delta}(q, t), |I| + 1), & \text{if } b = 1. \end{cases}$$

Here, $b = 1$ signals that we want to include k into the index set, and $b = 0$ signals that we do not want to. Observe that for two index sets I' and I with $I'|_{[k]} = I|_{[k]}$ it holds that $F_k^I = F_k^{I'}$. Thus for $k \geq 1$

$$\mathcal{C}_k = \{\mathrm{ext}(F, b) : F \in \mathcal{C}_{k-1}, b \in \{0, 1\}\}.$$

This computation takes time linear in $|\mathcal{C}_{k-1}|$ which is bounded by 2^{2^d}, so we can perform the computation in constant time. If there are more than one entries in $\mathcal{C}_k$ with the same F_k^I then we delete all but one with a maximum $|I|$. In the end we obtain $\mathcal{C}_n$. For each $(F, |I|) \in \mathcal{C}_n$, we evaluate F on φ_A. The maximum $|I|$ for which the F evaluates to true equals $d_{\mathrm{VC}}(S_n)$. The whole computation can be done in time $O(n)$. If we want to compute a maximum size shattered index set, rather than only its size, then we can for example in addition store for each formula the decision $b = 0$ or $b = 1$ and a reference from which formula it was derived. By the usual backtracking technique we can compute a maximum shattered index set then. □

4.7 Some S-SAT which is not NP-hard

S-SAT is NP-hard if and only if there is a polynomial Karp reduction φ from SAT to S-SAT. The reduction φ maps formulas to instances of S-SAT such that F is satisfiable if and only if $\varphi(F)$ is S-satisfiable. In this section we will prove that there is an exponential S such that no such reduction φ exists, provided that $\mathsf{P} \neq \mathsf{NP}$. An instance of S-SAT is a pair (F, V) with a formula F and $V \supseteq \mathrm{vbl}(F)$ an ordered set of variables. Two instances $(F_1, V_1), (F_2, V_2)$ are *equivalent*, denoted by $(F_1, V_1) \equiv (F_2, V_2)$, if $V_1 = V_2$ and F_1, F_2 agree on every assignment of the variables. Clearly, two equivalent instances are either both S-satisfiable or both are not S-satisfiable. We want to use a classical tool of complexity theory: diagonalization. For that we need a generalization of possible polynomial Karp reductions from SAT to S-SAT.

Definition 4.19. *A function φ mapping formulas to instances is called a* SAT-reduction *if, for all satisfiable formulas F and unsatisfiable formulas F', we have that $\varphi(F) \not\equiv \varphi(F')$. If there exists an algorithm which computes φ in polynomial time, then we say that it is a* polynomial SAT-reduction.

Consider for example the mapping φ which maps every satisfiable formula to $(\mathrm{true}, \emptyset)$ and every unsatisfiable formula to $(\mathrm{false}, \emptyset)$. This is a SAT-reduction but it is not polynomial (provided $\mathsf{NP} \neq \mathsf{P}$). We could also map satisfiable formulas to $(x, \{x\})$ and unsatisfiable formulas to $(x \wedge x, \{x\})$. However, this is *not* a SAT-reduction, since $(x, \{x\}) \equiv (x \wedge x, \{x\})$. If φ is not a SAT-reduction then there is a satisfiable formula F and an unsatisfiable formula F' such that $\varphi(F) \equiv \varphi(F')$.

Proposition 4.20. *Let S be fixed. If φ is a polynomial Karp reduction from* SAT *to* S-SAT*, then φ is a* SAT*-reduction as well.*

Proof. Assume for contradiction that φ is not a SAT-reduction. Then there is a satisfiable formula F and an unsatisfiable formula F' such that $\varphi(F) \equiv \varphi(F')$. Since φ is a Karp reduction, we have that $\varphi(F)$

is S-satisfiable and $\varphi(F')$ is not S-satisfiable. This is a contradiction, because two equivalent instances are either both S-satisfiable or both are not S-satisfiable. □

Lemma 4.21. *Provide that* $\mathsf{P} \neq \mathsf{NP}$, *then for every polynomial SAT-reduction* φ *and every* $n_0 \in \mathbb{N}_0$, *there exists a satisfiable formula* F *with* $n(\varphi(F)) \geq n_0$, *i.e., satisfiable formulas have arbitrarily large images under* φ.

Proof. For the sake of contradiction, suppose that there is some SAT-reduction φ and some n_0 such that $n(\varphi(F)) \leq n_0$ for all satisfiable F. Let $\mathcal{F}_0$ be the class of all instances that occur as an image of some satisfiable formula under φ. By assumption, all instances in $\mathcal{F}_0$ have not more than n_0 variables, implying that there are only a finite number of non-equivalent instances in $\mathcal{F}_0$. Clearly, F is satisfiable if and only if $\varphi(F) \equiv f \in \mathcal{F}_0$. For the finite language $\mathcal{F}_0$ we can store all elements in a look-up table and therefore $\mathcal{F}_0 \in \mathsf{P}$. Thus, SAT is in P, contradicting our assumption. □

Theorem 4.22. *Provided that* $\mathsf{P} \neq \mathsf{NP}$, *there is an* S *such that for all* $n \in \mathbb{N}_0$ *either* $|S_n| = 2^n$ *or* $|S_{n+1}| = 2^{n+1}$, *and* S-SAT *is not* NP-*hard.*

Proof. Let $\varphi_1, \varphi_2, \ldots$ be an enumeration of all polynomial SAT-reductions (there are countably many). For every $n_0 \in \mathbb{N}_0, i \in \mathbb{N}$ there exist by Lemma 4.21 a number $n(\varphi_i, n_0)$ and a satisfiable formula F such that $n(\varphi_i(F)) = n(\varphi_i, n_0) \geq n_0$. Define

$$n_1 := n(\varphi_1, 0),$$
$$n_{i+1} := n(\varphi_{i+1}, n_i + 2).$$

$$S_n := \begin{cases} \emptyset, & \text{if } n = n_i \text{ for some } i; \\ \{0,1\}^n, & \text{otherwise.} \end{cases}$$

Note that $n_{i+1} - n_i \geq 2$ and if $S_n = \emptyset$, then $|S_{n-1}| = 2^{n-1}$. Therefore, S is an exponential family with $\alpha = 2$ and $p(n) = n + 2$. Assume

4.7. Some S-SAT which is not NP-hard

for the sake of contradiction that S-SAT is NP-hard. Then there is a polynomial Karp reduction φ from SAT to S-SAT. By Proposition 4.20 φ is a polynomial SAT-reduction and therefore $\varphi = \varphi_i$ for some i. By construction, there is a satisfiable formula F such that $\varphi_i(F)$ has exactly n_i variables. But, S_{n_i} is empty, so $\varphi_i(F)$ is not S-satisfiable, hence φ_i is not a reduction, which is a contradiction. □

This is nice, but has the drawback that S might have gaps, i.e., not every level has exponential size. The problem above was that, in order to ensure that for the satisfiable formula F its image $\varphi(F)$ is not S-satisfiable, we set $S_n = \emptyset$ for $n = n(\varphi(F))$, creating a "gap" in S. For a SAT-reduction φ we define the functions φ^f and φ^v by $\varphi(F) = (\varphi^f(F), \varphi^v(F))$ for every formula F. Denote the set of all assignments of the variables $\varphi^v(F)$ which satisfy $\varphi^f(F)$ by $\text{sat}(\varphi(F))$. To assure that φ is not a allowed reduction, we could alternatively set $S_n = \{0,1\}^n \setminus \text{sat}(\varphi(F))$. Clearly, this suffices to ensure that $\varphi(F)$ is not S-satisfiable, preventing φ from being a reduction from SAT to S-SAT. If, in addition, $\text{sat}(\varphi(F))$ is small, $|S_n|$ will be exponential in n. Let us now first focus on what happens when it is never small.

Definition 4.23. *A* SAT-*reduction φ is called* sharp, *if there is some n_0 such that for all F with $n := n(\varphi(F)) \geq n_0$, the following two statements hold:*

(i) *F and $\varphi^f(F)$ are* SAT-*equivalent, that is, either both are satisfiable, or both are not;*

(ii) *if $\varphi^f(F)$ is satisfiable, then $|\text{sat}(\varphi(F))| > 2^{n-1}$.*

Any number x with $x/2^n > \epsilon > 0$ and $2^n - x$ being exponential would be fine as well in point (ii). The image of a sharp reduction consists of formulas with at most n_0 variables, unsatisfiable formulas, and formulas with a huge number of satisfying assignments.

Lemma 4.24. *If there is a polynomial sharp* SAT-*reduction φ, then* RP = NP.

Proof. We give a randomized algorithm for SAT with a bounded error probability. Similar to the proof of Lemma 4.21, let $\mathcal{F}_0$ contain all instances with less than n_0 variables which are the image of a satisfiable formula under φ. Again, this set is finite up to equivalence. There *exists* a randomized polynomial algorithm as follows: For all instances with less than n_0 we store in a look-up table whether their preimages under φ are satisfiable (recall that either all preimages are satisfiable or none). We compute satisfiability of some input formula F as follows: if $n(\varphi(F)) \leq n_0$, we simply check the look-up table, which can be done in constant time. Otherwise, either both F and $\varphi^f(F)$ are unsatisfiable, or both are satisfiable, but then $\text{sat}(\varphi(F))$ is huge. Choose x uniformly at random out of $\{0,1\}^n$ for $n = n(\varphi(F))$ and return *satisfiable* if x satisfies $\varphi^f(F)$ and *unsatisfiable* otherwise. If F is unsatisfiable, the algorithm always answers correctly, otherwise the answer is wrong with a probability $p \leq 1/2$. Thus SAT is in RP, and hence RP = NP. □

The contrapositive of Lemma 4.24 reads as follows: Provided that RP $\neq$ NP, no polynomial SAT-reduction φ is sharp, which means that for all φ, n_0, there exist $n = n(\varphi, n_0) \geq n_0$, $F = F(\varphi, n_0)$, such that $\varphi(F)$ has n variables and one of the following holds:

(i) F and $\varphi^f(F)$ are not SAT-equivalent, or
(ii) they are SAT-equivalent, $|\text{sat}(\varphi(F))| \leq 2^{n-1}$, and $\varphi^f(F)$ is satisfiable.

Theorem 4.25. *Provided that* RP $\neq$ NP, *there is an S with $|S_n| \geq 2^{n-1}$ for all n such that S-SAT is not NP-hard.*

Proof. Using the function $n(\varphi, n_0)$ and our sequence $\varphi_1, \varphi_2, \ldots$ of polynomial SAT-reductions, we define

$$n_1 := n(\varphi_1, 0), \qquad F_1 := F(\varphi_1, 0),$$
$$n_{i+1} := n(\varphi_{i+1}, n_i + 1), \qquad F_{i+1} := F(\varphi_{i+1}, n_i + 1).$$

The F_i are the formulas with n_i variables provided by the contrapositive of Lemma 4.24, and the n_i are all distinct. If case (i) above applies to

4.7. Some S-SAT which is not NP-hard

F_i, we say n_i is of type (i), if case (ii) applies, n_i is of type (ii). We define S by

$$S_n := \begin{cases} \{0,1\}^n \setminus \text{sat}(\varphi_i(F_i)) & \text{if n = n}_i \text{ is of type (ii);} \\ \{0,1\}^n & \text{otherwise.} \end{cases}$$

We claim that every φ fails to be a reduction from SAT to S-SAT. Take any φ_i. If n_i is of type (i), then F_i and $\varphi_i^f(F_i)$ are not SAT-equivalent, and since $S_{n_i} = \{0,1\}^{n_i}$, $\varphi_i(F_i)$ is S-satisfiable iff F_i is not satisfiable. Thus, φ is not a reduction from SAT to S-SAT. If n_i is of type (ii), then F_i and $\varphi_i^f(F_i)$ are both satisfiable, but $\varphi_i(F_i)$ is not S-satisfiable, since $S_{n_i} = \{0,1\}^{n_i} \setminus \text{sat}(\varphi_i(F_i))$. Hence φ_i fails also in this case. Finally, note that $|S_n| \geq 2^{n-1}$ for all n. □

We finish here by comparing our result of this section with a classical theorem by Ladner [62].

Theorem 4.26 (Ladner's Theorem [62]). *Provided* P $\neq$ NP*, there exists a language* $L \in$ NP $\setminus$ P *that is not* NP*-hard.*

Languages of this form are called NP-*intermediate* languages and it is an open question to construct any "natural" examples. It is not clear if we could use directly an NP-intermediate language L to construct an S such that S-SAT is not NP-hard. Moreover, there is no consideration in Ladner's Theorem how "dense" the language L is, where in Theorem 4.22 and Theorem 4.25 we have constructed *exponential* languages. However, it is not clear if there exists an exponential language S that is also in NP, such that the S-SAT problem is not NP-hard.

Bibliography

[1] S. Aaronson, G. Kuperberg, and C. Granade. Complexity Zoo. http://qwiki.stanford.edu/wiki/Complexity_Zoo.

[2] H. L. Abbott. Lower bounds for some Ramsey numbers. *Discrete Math.*, 2(4):289–293, 1972.

[3] E. Allender, M. Bauland, N. Immerman, H. Schnoor, and H. Vollmer. The complexity of satisfiability problems: Refining Schaefer's theorem. *Journal of Computer and System Sciences*, 75(4):245 – 254, 2009.

[4] N. Alon. Problems and results in extremal combinatorics. II. *Discrete Math.*, 308(19):4460–4472, 2008.

[5] N. Alon, R. Berke, K. Buchin, M. Buchin, P. Csorba, S. Shannigrahi, B. Speckmann, and P. Zumstein. Polychromatic colorings of plane graphs. *Discrete and Computational Geometry*, 42(3):421–442, October 2009. A preliminary version appeared in: *SoCG '08: Proceedings of the twenty-fourth annual symposium on Computational geometry*, pages 338–345.

[6] N. Alon, A. Krech, and T. Szabó. Turán's theorem in the hypercube. *SIAM J. Discrete Math.*, 21(1):66–72 (electronic), 2007.

[7] V. Arvind, J. Köbler, U. Schöning, and R. Schuler. If NP has polynomial-size circuits, then MA=AM. *Theor. Comput. Sci.*, 137(2):279–282, 1995.

[8] L. W. Beineke and A. J. Schwenk. On a bipartite form of the Ramsey problem. In *Proceedings of the Fifth British Combinatorial Conference (Univ. Aberdeen, Aberdeen, 1975)*, pages 17–22. Congressus Numerantium, No. XV, Winnipeg, Man., 1976. Utilitas Math.

[9] H. Bielak. Ramsey and 2-local Ramsey numbers for disjoint unions of cycles. *Discrete Math.*, 307(3-5):319–330, 2007.

[10] A. Biere, M. J. H. Heule, H. van Maaren, and T. Walsh, editors. *Handbook of Satisfiability*, volume 185 of *Frontiers in Artificial Intelligence and Applications*. IOS Press, February 2009.

[11] B. Bollobás. *Extremal graph theory*. Dover Publications Inc., Mineola, NY, 2004. Reprint of the 1978 original.

[12] P. Bose, D. Kirkpatrick, and Z. Li. Worst-case-optimal algorithms for guarding planar graphs and polyhedral surfaces. *Comput. Geom.*, 26(3):209–219, 2003.

[13] P. Bose, T. Shermer, G. Toussaint, and B. Zhu. Guarding polyhedral terrains. *Comput. Geom.*, 7(3):173–185, 1997.

[14] T. Brüggemann and W. Kern. An improved deterministic local search algorithm for 3-SAT. *Theor. Comput. Sci.*, 329(1-3):303–313, 2004.

[15] S. A. Burr. A survey of noncomplete Ramsey theory for graphs. In *Topics in graph theory (New York, 1977)*, volume 328 of *Ann. New York Acad. Sci.*, pages 58–75. New York Acad. Sci., New York, 1979.

[16] S. A. Burr. On the Ramsey numbers $r(G, nH)$ and $r(nG, nH)$ when n is large. *Discrete Math.*, 65(3):215–229, 1987.

[17] S. A. Burr, P. Erdős, R. J. Faudree, C. C. Rousseau, and R. H. Schelp. Ramsey-minimal graphs for star-forests. *Discrete Math.*, 33(3):227–237, 1981.

[18] S. A. Burr, P. Erdős, R. J. Faudree, and R. H. Schelp. A class of Ramsey-finite graphs. In *Proceedings of the Ninth Southeastern Conference on Combinatorics, Graph Theory, and Computing (Florida Atlantic Univ., Boca Raton, Fla., 1978)*, Congress. Numer., XXI, pages 171–180, Winnipeg, Man., 1978. Utilitas Math.

[19] S. A. Burr, P. Erdős, and L. Lovász. On graphs of Ramsey type. *Ars Combinatoria*, 1(1):167–190, 1976.

[20] S. A. Burr, P. Erdős, and J. H. Spencer. Ramsey theorems for multiple copies of graphs. *Trans. Amer. Math. Soc.*, 209:87–99, 1975.

[21] S. A. Burr, R. J. Faudree, and R. H. Schelp. On Ramsey-minimal graphs. In *Proceedings of the Eighth Southeastern Conference on Combinatorics, Graph Theory and Computing (Louisiana State Univ., Baton Rouge, La., 1977)*, pages 115–124. Congressus Numerantium, No. XIX, Winnipeg, Man., 1977. Utilitas Math.

[22] J.-Y. Cai. S2p$\subseteq$ZPPNP. *J. Comput. Syst. Sci.*, 73(1):25–35, 2007.

[23] L. Cai and J. A. Ellis. NP-completeness of edge-colouring some restricted graphs. *Discrete Applied Mathematics*, 30(1):15 – 27, 1991.

[24] V. Chvátal. A combinatorial theorem in plane geometry. *J. Combinatorial Theory Ser. B*, 18:39–41, 1975.

[25] D. Conlon. A new upper bound for diagonal Ramsey numbers. *Ann. of Math.*, 170(2):941–960, 2009.

[26] S. A. Cook. The complexity of theorem-proving procedures. In *STOC '71: Proceedings of the third annual ACM symposium on Theory of computing*, pages 151–158, New York, NY, USA, 1971. ACM.

[27] J. Cooper. Josh Cooper's Math Pages: Comb. problems I like. http://www.math.sc.edu/~cooper/combprob.html.

[28] E. Dantsin, A. Goerdt, E. Hirsch, R. Kannan, J. Kleinberg, C. Papadimitriou, O. Raghavan, and U. Schöning. A deterministic $(2 - 2/(k+1))^n$ algorithm for k-SAT based on local search. In *Theoretical Computer Science 289*, pages 69–83, 2002.

[29] R. Diestel. *Graph theory*, volume 173 of *Graduate Texts in Mathematics*. Springer-Verlag, Berlin, third edition, 2005.

[30] D. Dimitrov, E. Horev, and R. Krakovski. Polychromatic colorings of rectangular partitions. *Discrete Mathematics*, 309(9):2957 – 2960, 2009.

[31] Y. Dinitz, M. Katz, and R. Krakovski. Guarding rectangular partitions. In *Abstracts 23rd European Workshop on Computational Geometry*, pages 30–33, 2007.

[32] P. Erdős. Some remarks on the theory of graphs. *Bull. Amer. Math. Soc.*, 53:292–294, 1947.

[33] P. Erdős, R. J. Faudree, C. C. Rousseau, and R. H. Schelp. The size Ramsey number. *Periodica Mathematica Hungarica*, 9(2-2):145–161, 1978.

[34] P. Erdős and G. Szekeres. A combinatorial problem in geometry. *Compositio Math.*, 2:463–470, 1935.

[35] G. Exoo. A lower bound for $R(5,5)$. *J. Graph Theory*, 13(1):97–98, 1989.

[36] R. Faudree. Ramsey minimal graphs for forests. *Ars Combin.*, 31:117–124, 1991.

[37] R. J. Faudree and R. H. Schelp. A survey of results on the size Ramsey number. In *Paul Erdős and his mathematics, II (Budapest, 1999)*, volume 11 of *Bolyai Soc. Math. Stud.*, pages 291–309. János Bolyai Math. Soc., Budapest, 2002.

[38] S. Fisk. A short proof of Chvátal's watchman theorem. *J. Combin. Theory Ser. B*, 24(3):374, 1978.

[39] J. Folkman. Graphs with monochromatic complete subgraphs in every edge coloring. *SIAM J. Appl. Math.*, 18:19–24, 1970.

[40] J. Fox and K. Lin. The minimum degree of Ramsey-minimal graphs. *Journal of Graph Theory*, 54(2):167–177, 2006.

[41] H. Gebauer. Disproof of the neighborhood conjecture with implications to sat. In *ESA*, pages 764–775, 2009.

[42] L. Gerencsér and A. Gyárfás. On Ramsey-type problems. *Ann. Univ. Sci. Budapest. Eötvös Sect. Math.*, 10:167–170, 1967.

[43] S. Ginsburg. *The mathematical theory of context-free languages.* McGraw-Hill, Inc., New York, NY, USA, 1966.

[44] R. L. Graham, B. L. Rothschild, and J. H. Spencer. *Ramsey theory.* Wiley-Interscience Series in Discrete Mathematics and Optimization. John Wiley & Sons Inc., New York, second edition, 1990. A Wiley-Interscience Publication.

[45] R. E. Greenwood and A. M. Gleason. Combinatorial relations and chromatic graphs. *Canad. J. Math.*, 7:1–7, 1955.

[46] B. Guenin. Packing t-joins and edge colouring in planar graphs. 2003.

[47] R. P. Gupta. On the chromatic index and the cover index of a multigraph. In *Theory and applications of graphs (Proc. Internat. Conf., Western Mich. Univ., Kalamazoo, Mich., 1976)*, pages 204–215. Lecture Notes in Math., Vol. 642. Springer, Berlin, 1978.

[48] S. L. Hakimi and O. Kariv. A generalization of edge-coloring in graphs. *J. Graph Theory*, 10(2):139–154, 1986.

[49] P. J. Heawood. On the four-color map theorem. *Quarterly Journal of Pure and Applied Mathematics*, 29:270–285, 1898.

[50] F. Hoffmann and K. Kriegel. A graph-coloring result and its consequences for polygon-guarding problems. *SIAM J. Discrete Math.*, 9(2):210–224, 1996.

[51] I. Holyer. The NP-completeness of edge-coloring. *SIAM J. Comput.*, 10(4):718–720, 1981.

[52] E. Horev, M. J. Katz, and R. Krakovski. Polychromatic coloring of cubic bipartite plane graphs, 2009. submitted.

[53] E. Horev and R. Krakovski. Polychromatic colorings of bounded degree plane graphs. *Journal of Graph Theory*, 60(4):269–283, 2009.

[54] T. R. Jensen and B. Toft. *Graph coloring problems*. Wiley-Interscience Series in Discrete Mathematics and Optimization. John Wiley & Sons Inc., New York, 1995. A Wiley-Interscience Publication.

[55] S. Jukna. *Extremal Combinatorics*. Springer-Verlag, 2001.

[56] R. M. Karp and R. J. Lipton. Some connections between nonuniform and uniform complexity classes. In *Enseign. Math. 28*, pages 191–201, 1982.

[57] A. B. Kempe. On the geographical problem of the four colours. *American Journal of Mathematics*, 2(3):193–200, 1879.

[58] B. Keszegh. Polychromatic colorings of n-dimensional guillotine-partitions. In *COCOON*, pages 110–118, 2008.

[59] J. H. Kim. The Ramsey number $r(3, t)$ has order of magnitude $t^2/\log t$. *Random Struct. Algorithms*, 7(3):173–207, 1995.

[60] J. Köbler and O. Watanabe. New collapse consequences of NP having small circuits. *SIAM J. Comput.*, 28(1):311–324, 1999.

[61] J. Kratochvíl, P. Savický, and Z. Tuza. One more occurrence of variables makes satisfiability jump from trivial to NP-complete. *SIAM Journal of Computing*, 22(1):203–210, 1993.

[62] R. E. Ladner. On the structure of polynomial time reducibility. *J. ACM*, 22(1):155–171, 1975.

[63] L. A. Levin. Universal search problems. *Problemy Peredači Informacii*, 9(3):265–266, 1973.

[64] L. Lovász and M. D. Plummer. *Matching theory*, volume 121 of *North-Holland Mathematics Studies*. North-Holland Publishing Co., Amsterdam, 1986. Annals of Discrete Mathematics, 29.

[65] B. D. McKay and S. P. Radziszowski. Subgraph counting identities and Ramsey numbers. *J. Combin. Theory Ser. B*, 69(2):193–209, 1997.

[66] B. Mohar and R. Škrekovski. The Grötzsch theorem for the hypergraph of maximal cliques. *Electron. J. Combin.*, 6:Research Paper 26, 13 pp. (electronic), 1999.

[67] B. M. E. Moret. Planar NAE3SAT is in P. *SIGACT News*, 19(2):51–54, 1988.

[68] R. A. Moser. A constructive proof of the Lovász local lemma. In *STOC*, pages 343–350, 2009.

[69] J. Nešetřil and V. Rödl. The structure of critical Ramsey graphs. *Acta Math. Acad. Sci. Hungar.*, 32(3-4):295–300, 1978.

[70] J. Nešetřil and V. Rödl. Simple proof of the existence of restricted Ramsey graphs by means of a partite construction. *Combinatorica*, 1(2):199–202, 1981.

[71] D. Offner. Polychromatic colorings of subcubes of the hypercube. *SIAM J. Discrete Math.*, 22(2):450–454, 2008.

[72] C. H. Papadimitriou. *Computational Complexity*. Addison Wesley, 1994.

[73] C. H. Papadimitriou and M. Yannakakis. On limited nondeterminism and the complexity of the V-C dimension. *J. Comput. Syst. Sci.*, 53(2):161–170, 1996.

[74] S. Radziszowski. Small Ramsey numbers. *Electronic Journal of Combinatorics*, 2006.

[75] F. P. Ramsey. On a problem of formal logic. *Proc. London Math. Soc.*, s2-30(1):264–286, 1930.

[76] V. Rödl and A. Ruciński. Threshold functions for Ramsey properties. *J. Amer. Math. Soc.*, 8(4):917–942, 1995.

[77] N. Sauer. On the density of families of sets. In *J. Comb. Theory, Ser. (A)*, volume 13, pages 145–147, 1973.

[78] T. J. Schaefer. The complexity of satisfiability problems. In *STOC '78: Proceedings of the tenth annual ACM symposium on Theory of computing*, pages 216–226, New York, NY, USA, 1978. ACM.

[79] D. Scheder. Guided search and a faster deterministic algorithm for 3-SAT. In *LATIN*, pages 60–71, 2008.

[80] D. Scheder and P. Zumstein. Satisfiability with exponential families. In *SAT*, pages 148–158, 2007.

[81] D. Scheder and P. Zumstein. How many conflicts does it need to be unsatisfiable? In *SAT*, pages 246–256, 2008.

[82] P. D. Seymour. On multi-colourings of cubic graphs, and conjectures of fulkerson and tutte. *Proc. London Math. Soc*, 3:423–460, 1979.

[83] S. Shelah. A combinatorial problem; stability and order for models and theories in infinitary languages. *Pacific J. Math.*, 41:247–261, 1972.

[84] A. Soifer. *The mathematical coloring book*. Springer, New York, 2009. Mathematics of coloring and the colorful life of its creators, With forewords by Branko Grünbaum, Peter D. Johnson, Jr. and Cecil Rousseau.

[85] J. Spencer. Ramsey's theorem—a new lower bound. *J. Combinatorial Theory Ser. A*, 18:108–115, 1975.

[86] R. Steinberg. The state of the three color problem. In *Quo vadis, graph theory?*, volume 55 of *Ann. Discrete Math.*, pages 211–248. North-Holland, Amsterdam, 1993.

[87] L. Stockmeyer. Planar 3-colorability is polynomial complete. *SIGACT News*, 5(3):19–25, 1973.

[88] T. Szabó, P. Zumstein, and S. Zürcher. On the minimum degree of Ramsey-minimal graphs. to appear.

[89] V. N. Vapnik and A. Y. Chervonenkis. On the uniform convergence of relative frequencies of events to their probabilities. *Theory Prob. Appl.*, 16:264–280, 1971.

[90] H. Vollmer. *Introduction to circuit complexity*. Texts in Theoretical Computer Science. An EATCS Series. Springer-Verlag, Berlin, 1999. A uniform approach.

[91] D. B. West. *Introduction to Graph Theory*. Prentice Hall, second edition, 2001.

Die VDM Verlagsservicegesellschaft sucht für wissenschaftliche Verlage abgeschlossene und herausragende

Dissertationen, Habilitationen, Diplomarbeiten, Master Theses, Magisterarbeiten usw.

für die kostenlose Publikation als Fachbuch.

Sie verfügen über eine Arbeit, die hohen inhaltlichen und formalen Ansprüchen genügt, und haben Interesse an einer honorarvergüteten Publikation?

Dann senden Sie bitte erste Informationen über sich und Ihre Arbeit per Email an *info@vdm-vsg.de*.

Sie erhalten kurzfristig unser Feedback!

VDM Verlagsservicegesellschaft mbH
Dudweiler Landstr. 99
D - 66123 Saarbrücken
www.vdm-vsg.de

Telefon +49 681 3720 174
Fax +49 681 3720 1749

Die VDM Verlagsservicegesellschaft mbH vertritt

Printed by Books on Demand GmbH, Norderstedt / Germany